REAL Success were integral t
Weekly Dairy Farm of the Ye ... and our work together began
with a Team Dynamics session and profiling all our staff. It remains a vital part of our understanding of our team and each other. This story is typical of many farming families and teams, and the lessons cleverly woven into the narrative will help them build a stronger and more successful team.

Geoff Sayers - farm owner, The Carswell Group

We've worked with Paul and his team for several years and recognise many of the pressures and challenges outlined in this book. Establishing a deeper understanding of each other, whether that is a family member or colleague, has been the foundation of our development.

Austin Knowles - farm owner, Hollings Hill Farm

When we first engaged REAL Success to help us, we started by profiling the team and continue to use our understanding of the different types. We also use it when recruiting to ensure new employees fit into our team. This story shows the importance of recognising the distinct characteristics of all those who work with you, and we've found it an invaluable tool to help us.

Mike King – farm owner, Two Pools Farm

I recognise many of the issues raised in this book, and with a large and diverse team, our managers know that understanding individual personalities is critical to our ongoing success.

David Craven – farm manager, Grosvenor Farms

When I completed a VITA Profile, I couldn't believe how accurately it reflected who I am and how I work. It's also helped me understand how to work more effectively with my colleagues.

James Major – farm manager

Profiling has helped us understand how to get the best out of our staff, and we continue to use the lessons that profiling gives us. This book is a simple introduction to some basic concepts of understanding your team and how to get the best from them.

Ben Atkinson – farm owner, Atkinson Farming

Our family and farm team have worked together more happily since we completed the Team Dynamics workshop. We also understand our differences and similarities and how this helps us be more patient with each other. This story resonated with me, and I'm sure it will help anyone trying to find their way on their family farm and with their team.

Charlotte West – family member, Stocken Farm

When REAL Success delivered a Team Dynamics workshop at EDF Congress in Denmark, the workshop was ranked the best at the congress. I'm delighted that this story covers some of the concepts in the talk and I'm sure it will help our farmers of all nationalities to understand themselves and their teams even more.

Judit Kuehl – EDF Congress

This story resonated strongly with me as I understand the pressure that Peter placed upon himself. When I discovered my personality style, I realised I needed to allow others to help me, which transformed my business and personal life.

Tom Foot – farm owner, Longlands Farm

We are a farming family with an extensive team, so many of the issues raised in *Happy Team, Happy Farm* are issues we have faced as a family. Understanding our personality styles has hugely helped us manage our team more effectively.

Matt Pilkington – farm owner, Pilkington Farms

As a large farming family with vastly different personalities, understanding how we may each approach situations on the farm has helped us work together more effectively. I'm certain this book will help other farm owners and their staff to appreciate their differences.

Selena Richards – farm owner, CP Richards & Son

PAUL HARRIS

HAPPY TEAM HAPPY FARM

A *story* of family, farming and the real value of difference and diversity

Happy Team, Happy Farm

ISBN 978-1-912300-66-2
eISBN 978-1-912300-67-9

Published in 2023 by Right Book Press

A CIP record of this book is available from the British Library.

Happy Team, Happy Farm is a work of fiction. Any resemblance to actual events or persons, living or dead, is entirely coincidental. Certain locations and geographical references are real but any characters, situations or businesses referred to are wholly imaginary except for REAL Success, which is an established business based in the UK.

Contents

Foreword

The farming industry is facing unprecedented change, from how the government supports us to changes in the environment, regenerative practices and new science and technology. Our industry has never faced so much change at the same time. However, one constant aspect is the need to manage, develop and retain people. Creating a positive working environment where staff (and family) want to stay and build their careers is our most critical challenge. Yet this aspect of our leadership and management skills has been neglected for a long time.

Since we began working with Paul and his team at REAL Success, my eyes have been opened to the challenges and opportunities that developing my leadership skills can give to my business. Recognising that different people in my team need to be managed differently was a revelation. While this may seem obvious, understanding how to manage my team differently was a lot of hard work. The VITA Profiling tool enabled us to develop our management and leadership style to meet the needs of each person in our team while also recognising that the leadership team needed to learn new skills. Our journey to becoming better managers and leaders began with a simple understanding of profiling, gently covered in this book.

While the story is based on a farming family with a small staff team, the messages and situations covered will resonate with farmers of all types and non-farming businesses too. I have found the journey of discovering how to get the best from my team both challenging and immensely rewarding. I encourage you to read the story with an open mind and then decide to understand your team and, if appropriate, your family. It may be the making of you and your business.

Rupert Major, farm owner, Major Farming Ltd

Author's note

This book has been written to help the farming community manage their people more effectively and hopefully make a lasting and positive difference to the agricultural industry. However, all the messages, principles and lessons contained within the story apply to most businesses – whichever market they may be operating in.

People skills are often the most overlooked element of a farming business, with far more emphasis placed on developing skills with livestock, land and machinery. There are many large and small consultancy firms that support the agricultural sector with technical and business advice while leadership, management and people skills are often viewed as soft skills and not as important to the performance of the business. However, in all the exit interviews that my company, REAL Success, undertakes as part of our recruitment services to farmers, the single biggest reason people cite as their decision to leave a business is the way in which someone (often the boss) speaks to them. This book attempts to address this imbalance of knowledge by offering a starting point.

Understanding how your team prefer to communicate and behave is the foundation of all further leadership and management development. If you are unable to understand your team and adjust your communication style to get the best from them, no amount of leadership training will work. Knowing your team members are all different and need to be spoken to and managed differently can transform the performance of your team and the profitability of your business.

In my experience, the scenario in this story is incredibly common. A family that doesn't understand their differences and how to work with each other can end up fracturing. A team that doesn't understand why their colleagues may behave differently can end up in conflict, reducing morale and diminishing performance. Some staff will simply leave the business with the resulting inconvenience and associated costs.

The farming community in the UK is also undiversified. Dominated by white, British males with a very limited ethnic or LGBTQIA+ representation, this story also offers a perspective on how the industry could respond.

As you read the story, I urge you to consider your own business, your own prejudices and your own personality. In doing so, this book will be more alive and relevant to your own situation.

Together we can modernise and improve the reputation and working practices within the farming industry in the UK and beyond. I look forward to hearing your own stories.

Paul Harris, January 2023

This book is dedicated to my father, who believed that family is our most important gift.

'In time of test, family is best.'
– Burmese proverb

Scenario

Our story is set around Wilson Farm, a fictional mixed farm located near a small village in the county of Devon, England. With 800 acres of mixed pasture, the farm has just over 300 Friesian cows that live outside for most of the year as well as 400 Longwool sheep, a few chickens and crops grown as food for the cows, with surplus sold to local farmers. With a busy arable and livestock operation, the farm is one of the oldest and most well established in the area with, at one time, three generations of the same family working together on the farm.

The business was initially set up by Arthur and Millie Wilson, who bought a few acres of land to rear beef livestock and grow crops to develop a self-sustaining lifestyle. Arthur and Millie later moved into dairy farming, initially in a very low-key way, before investing in a milking parlour and housing for the animals.

Their son, Peter, now 52, started to work on the farm after leaving school and when he married Mary, 51, they became more involved, eventually taking over the running of the farm when Millie passed away peacefully in her sleep. Arthur continued to work around the farm as it gave him a purpose and kept him busy.

Peter and Mary have two children – Charlotte and Martyn. Charlotte, 26, is strong willed and ambitious. During her school days she was the captain of her hockey team and head girl of her high school. She is driven to succeed and was determined to work on the farm as soon as she finished her degree in agriculture. Having strong views, she set about changing calving routines and the management of the youngstock – with both the cows and sheep on the farm. While her love is for the calves, she enjoys helping with the sheep, particularly in lambing season.

Martyn is 22 and very different to his sister. His reading and writing are poor for his age and while at school he was often accused of being disruptive and lacking in concentration. Martyn simply wasn't interested in schoolwork. He spent every moment he could working on the farm, helping with milking the cows and herding the sheep. He is practical with bundles of energy and enthusiasm.

The family are supported by a mix of long-serving and recently appointed staff: Billy is the oldest member of the team and had worked for Arthur before Peter took over. While he is beyond retirement age, he still loves to come in every day to milk and look after his 'girls', as he affectionately refers to the cows. Andrew is the next longest serving member of the team and is responsible for the flock of Longwool sheep. He has never married and often suggests he is actually married to the farm. The other staff members are Amy, responsible for the calf rearing; Marek, who milks the cows; and George, the youngest member of the team, who carries out most of the tractor and general farm work.

Peter takes the lead on the day to day running of the farm while Mary focuses on paperwork – a task Peter finds frustrating and time consuming. Mary manages the accounts, the suppliers and the passport paperwork for the animals, while Peter manages the staff, the animals and the land. Between the two of them, they have found a way to create a profitable farming enterprise without taking too many risks. While others in the industry expanded their herds and dairy or arable enterprises to large production units, Peter and Mary always tried to keep a balance between creating a profitable and sustainable farm while also creating strong family bonds and relationships. The fact that both Charlotte and Martyn wanted to work on the family farm was proof to Peter and Mary that they'd got the balance about right. But the prospect of a dramatic change is rocking the family...

THE STORY

PART ONE

Family Tensions Revealed

1

'If that's how you feel, I'm leaving.' Martyn was furious and knew he could no longer work with his father. He'd tried to get him to listen to his new ideas but each time he'd proposed a change on the farm, his dad shut him down.

'You're too hot headed,' said Peter. 'You just want to change everything. Your grandad built this farm and I've taken care of everything since he died. Just because you've been to a discussion group and listened to a bunch of young farmers who think they know it all doesn't make your ideas workable on this farm. Why can't you just follow what we've always done?'

'Because nothing ever changes around here, Dad,' replied Martyn. 'Just look at the tractor. It's rusty, old and worn out – just like the farm. We need to modernise and try out new things; otherwise the farm will go backwards.'

'But what we do here works, Martyn. Sticking to what we know gives us security for the future. Your ideas threaten that and I'm just not ready to change everything on a whim.'

The atmosphere was cold, fractious and so far from where they'd been a few years ago. When Martyn left agricultural college, he was excited about running the family farm and couldn't wait to get stuck into the daily grind of milking cows, driving tractors and looking after the sheep. It was a busy farm but one with so much potential.

Peter had been excited about Martyn joining the family business. Charlotte, who was three years older than Martyn, was already working on the farm and taking charge of the calving operations and rearing youngstock. She was a little blunt and ruffled feathers among the staff but at least she was sticking to the plan. Martyn just wanted to change everything and it was causing conflict among the team.

'I think it's best if I go to work on another farm, Dad, where they're prepared to consider new ideas and I can grow my career. I'm in danger of stagnating here as you're never going to consider new methods or approaches.'

Peter was exhausted by the constant battles with his son. He loved him dearly and didn't want to see him leave the farm simply because of different ideas on how it should be run. But he also knew his relationship with his son was now at breaking point. They would fall out every day and now he would dread bumping into him around the farm. Martyn simply didn't respect his father anymore and the language he used towards him shocked other members of staff. Billy had previously expressed concern about Martyn's behaviour and, as one of the longest-serving members of the team, Billy's views mattered to Peter.

Martyn was equally frustrated and upset with the situation. He'd hoped to be running the farm by now but his father was just getting in the way of everything he wanted to do. He loved and respected his father and wanted their relationship to work but he'd often avoid him. He'd lost his temper a few times in front of the staff and Billy suggested he needed to mind his language. Martyn respected Billy and considered his advice.

'Listen, Dad – I don't want to upset the applecart here, I just want to do what's best for you. Since we lost Grandad, you've been working crazy hours and I know I could make things easier for you if you'd let me. I don't want to argue every day; I just want to make the farm a smart and forward-thinking place to work. Maybe the farm isn't ready for me or my ideas just yet so perhaps I do need to get away for a while for both our sakes?'

Peter put his hand on Martyn's shoulder. 'Your mum will be devastated if you leave the farm but perhaps it might be for the best for both of us and the rest of the team if you worked elsewhere for a while. The door will always be open for you to come back, Martyn – you know that.'

Martyn could feel emotion rising within him but didn't want to cry in front of his dad. 'Yes, I think you're right, Dad. I'll put together my CV and start looking for a job. I know Howard down the road was looking for a decent herdsman and I've heard of a chap who runs a recruitment service for farmers, so I'll give him a call too. It'll be sad to leave the farm but I think it's for the best.'

Martyn turned away from his dad and began to walk back towards the milking parlour to check how the team were getting on. As he walked, he wiped the tears from his eyes with the edge of his sleeve.

Peter watched Martyn walk away from him towards the parlour. His heart ached for the boy, who had so much to offer but so much to learn. He wondered where it had gone wrong between them. He turned away from his son and began to walk back towards the farmhouse to break the news to Mary. He knew she was going to blame him for Martyn's decision to leave the farm. As he walked, he wondered what his own father would have done if they'd struggled to get along.

Peter's dad Arthur had passed away nearly a year ago, after a short illness. Peter was still dealing with the loss while also realising just how much knowledge and experience had been lost from the farm with his passing. His dad's calm demeanour had been replaced with the more volatile energies of his grandchildren. Staff were grumbling and Peter wondered if all their hard work over the past 20 years was going to come crumbling down around them.

When Peter told Mary about his conversation with Martyn, Mary was utterly deflated. She felt she'd failed her husband and her son as she wasn't able to get them to resolve their conflict. She'd found it hard to watch the collapse of their relationship and the prospect of Martyn working elsewhere was an even heavier load for her to bear. Peter was more pragmatic about the situation.

'There's no point crying over spilt milk,' he said. 'We've just got to accept that Martyn wants to work elsewhere and there's nothing we can do about it.'

'You're just too stubborn, Peter Wilson, that's your problem,' said Mary. 'If you'd just considered a few of his ideas, he might have been happy to stay. I can't believe everything he suggested would have been a disaster.'

'But you know what he's like,' said Peter. 'He never sticks at anything. He just flits from one idea to the next so if I'd listened to him, the farm would have been turned upside down by now.'

Peter swirled the tea around in his mug, wondering what to say to Mary to reassure her. They sat opposite each other at the kitchen table where so many lively family discussions had taken place. Mary broke the silence.

'OK, love. I think we need some help to sort this mess out. I was talking to Anna the other day and they use a consultant who's supposed to be a bit of a wizard when it comes to sorting out people on farms. I'm not sure who he is or where he's based but I'll call Anna to get his details and give him a call to see if he can help us. I don't want my family falling apart or the business failing just because we can't seem to work with each other. There must be a better way.'

Peter pushed his mug towards Mary. 'You do whatever you think is best, Mary. I'm off to move fences.'

Peter went into the utility room to find his boots and overalls. Mary heard him leave the house and watched him walk across the yard towards the milking parlour. She wondered if his shoulders were a little more stooped than usual. She thought about how much she loved him but her heart craved peace and harmony rather than the bitterness and conflict of the past few months.

'Perhaps it would be better if Peter and Martyn weren't working together,' she thought. But it would mean finding a new herdsperson to replace Martyn on the farm and more upheaval for the team. She knew Peter hated interviews with staff and would often just recruit the first person who showed an interest in the farm. Mary wanted this next appointment to be right, so she picked up her mobile phone and dialled Anna.

'Hi Anna, it's Mary. How are you?'

'I'm really well, thanks. How are you?'

Mary decided to be honest with Anna as she'd known her for a long time. 'Well, I'm not great really. Martyn has decided to leave the farm because he can't work with Peter and it's such a shame. But you know what Peter's like – he's so stubborn. He's not prepared to work things out with Martyn so now I'm losing my son from the farm. It's so close to losing Gramps too; it's hit me quite hard.'

'Oh gosh, I'm sorry to hear this, Mary. You need to speak to the Chairman. You know, the chap I told you about?'

'The Chairman? Is that what people call him? What's his name, Anna?'

'To be honest, Mary, I don't know. Charlie always refers to him simply as the Chairman. He's used his advice for years and it completely transformed our team and relationships. And you know how much trouble we were having trying to find and keep staff.'

'Yes, I do remember and now I'm jealous of your settled team and the way your three children work with you in the business. Can you give me his number? I think I'll give him a call.'

'Sure,' said Anna. 'I'll text it to you now. But remember what Charlie says about him. He's not your average farm consultant or business guru. He asks all sorts of questions before he'll offer to help, so be ready to be cross-examined!'

Mary laughed and it lightened her mood. 'Thanks, Anna. I'll call him now.'

'Get ready to be open and brave, Mary – you won't regret it, I promise.'

Mary wondered why she would need to be brave. She stared at her phone wondering who the Chairman was and how he could help their farm. There was only one way to find out. She dialled the number and it began to ring. Mary shivered as if someone had just walked silently into the room. But there was no one around. After a few rings, a gentle voice answered. She was through to the Chairman…

2

'I'm not sure, Mary,' said Peter, pushing his dinner around his plate. 'You remember when we took on Charlie, don't you? We hoped he'd work towards running the farm one day but it was a disaster. He just upset everyone and I nearly lost Andrew. I'm not sure we need a farm manager anyway. I can cope with the workload... most of the time.'

But Peter knew he hadn't been coping for a while. In the months before his father became ill, they'd spent many hours together discussing the future of the farm. When the day finally came and Arthur passed away, Peter wasn't prepared at all.

'I was hoping Martyn would pick up the slack but I think I can cover his workload. I don't think we need a farm manager, Mary.'

'You're a stubborn fool,' said Mary, barely able to hide her frustration with her exhausted husband. 'You're even too tired to eat the food I prepare for you. And you're losing weight too. If you're not careful, you'll end up having a heart attack or a nervous breakdown, or both!'

Peter *was* finding it difficult to enjoy his food lately. He had constant indigestion and would often skip meals and keep working. Then he'd snack in the evening and spend his nights tossing and turning in bed.

'You worry too much, Mary. I'll be fine, you'll see.'

But Peter knew he wasn't convincing Mary or doing a very good job of convincing himself. He decided to try a different tack.

'We can't afford a farm manager anyway. We need to tighten our belts and we're already spending too much on labour. We need more cows, not more labour.'

Mary experienced a heady mix of frustration, worry and anger. Peter had become indoctrinated by John Hemmings, their farm business consultant, who was constantly encouraging Peter to buy more livestock, build more sheds for the animals and put the family into higher and higher levels of debt – something Gramps was always against. 'Aim for less' was a phrase Gramps used when other farmers were encouraging him to expand. His philosophy never lacked ambition but was rooted in recognising what they already had rather than constantly believing a bigger farm would bring bigger rewards. It was often the complete opposite for those who expanded their farms when they became exhausted and unable to take any time off.

Mary couldn't understand why Peter still listened so closely to John Hemmings, who seemed happy to spend his clients' money and enjoy his salary without any risk to his own lifestyle. Hemmings worked for a large farm consultancy that provided a range of business and technical advice but often seemed to ignore the impact of their advice on the family or people who worked for their clients. It was a gap that Mary had frequently pointed out to Peter while he drove the family further into debt and, more worrying for Mary, towards greater levels of stress.

'More cows aren't the answer, Peter. Nor is you running yourself into the grave right behind your father.' Mary's voice tailed off and she realised she'd dealt Peter a painful blow. The sound of knives and forks scraping on the plates filled the awkward silence that ensued.

'I'm sorry, love. I didn't mean to bring Gramps into this but you've got to recognise that with Martyn leaving too, we've got to do something different or else you're going to make yourself ill.'

Peter continued to push his food around his plate. Since he'd lost his father, he'd struggled to get his head around the complexities of his farming business. He longed for the chance to talk it through with his father. 'Dad would know what to do,' he thought.

'I'll have a chat with John tomorrow and see what he thinks,' said Peter.

Mary knew what John would say even before Peter called him. 'But won't you even consider talking to the Chairman instead?' she asked, hoping she could break through her husband's resistance.

'Well, who is this Chairman bloke anyway? You don't even know his name. Why would I listen to a chap Anna has recommended, you've not met and I'm now supposed to believe is the answer to all my troubles because he's some sort of business guru? I think I'll stick with my tried and tested people, thank you.'

Mary's shoulders slumped even further but she wasn't beaten yet. 'OK, Peter, if you won't meet him, I will. We've got to find a different solution for this farm and this family. You certainly aren't going to get off your backside and help yourself, so I'll have to do it for you. No, hang on. This isn't about Gramps or John Hemmings or anyone else. It's about you, me, Charlotte and Martyn. I'm going to talk to the Chairman again and see what more he has to say. Anna said her life and family have never been better and they put it all down to listening to the advice and guidance of the Chairman.'

'Do what you will, Mary, but I'm still going to speak to John tomorrow.'

Mary scraped the remains of her dinner into the waste bin and recalled her first conversation with the Chairman. It was as if he'd been listening to conversations they'd been having around the meal table or been present at their occasional staff meetings. He'd suggested to her that many farming families and teams were faced with similar challenges to those at Wilson Farm and that Mary shouldn't get too downhearted. He'd also told Mary that her first step to finding clarity in her thoughts was to 'find the silence'. When she'd asked what he meant, he told her to find space between her thoughts as too much busyness created a blurry and dark view of what was happening around her. He'd suggested she needed to create time to think and to walk around the farm alone with her thoughts. He felt Peter needed to do this as well. Mary laughed at the thought of Peter taking any time to think. He was always 'too busy' to take any time off and when he did take a day away from the farm, it was often to visit another farm, a show or to wander aimlessly around an exhibition. She knew many other farmers did the same things and rarely took time to just be with their businesses, families or partners.

The Chairman also talked to her about the need to recruit slowly. He suggested that many farmers didn't take their time to step back and consider what sort of person they needed or what they would do on their farm, instead just employing the first person who applied for a role on their farm. He referred to this as 'recruiting in haste and regretting at leisure'.

Mary liked what the Chairman had said. There was something about his voice that calmed her nerves. He'd simply listened, asked questions and made suggestions. He'd said they needed to step back from the business and consider what they truly wanted from their farm, their lives and each other. Mary realised she didn't know why they were doing what they were doing anymore. The fun was no longer there and watching her husband and son break apart her family, let alone the impact it could have on the business, was a huge worry and disappointment.

Peter got up from the table and put his arm around Mary's waist as she busied herself washing the plates and pans. 'I'm sorry, love. I know you mean well. Yes, I'm a stubborn old fool at times but one who still loves you, this farm and my family. We'll pull through, you'll see.'

Mary wasn't sure anymore. She couldn't see how John Hemmings, Tom Storey (the farm vet) or Jim Stanley (the accountant) were going to save their family. Despite the collection of advisors and consultants providing business support, she knew the future, if they were to have one, was no longer about the cows, sheep or land. It was about people. It was clear to her, for the first time in a long while, that if the farm were to survive, they needed the right people on the farm. And she wanted those people to be happy, including her family.

'I'm going to call the Chairman again and see if I can arrange to see him,' she said defiantly. 'I think he might be our best hope of keeping this family and this business and you, my dear stubborn husband, alive!'

3

It was 4 am and Martyn was awake and angry – not with his dad but the situation he was in. All his life he'd dreamed of running the farm and now he was looking for another job, all because his dad wouldn't accept any of his ideas and they couldn't agree on a way forward for the farm.

He pulled back the bedcovers and got up. Dressing quickly, he headed downstairs into the kitchen to make a quick coffee before heading out to join the milking team.

'Morning, son,' said Peter.

'Morning, Dad.'

Peter held a cup in his hand and Martyn knew he'd already been out to get the cows in for milking. He sat down at the kitchen table, concerned that his father was already working.

'You're up bright and breezy this morning, Dad. Wasn't George supposed to be getting the girls in this morning? How come you've been out already?'

'George texted me at 2 am and said he wasn't going to make it in this morning. I reckon he'd consumed a skinful again and just didn't want to drive with a hangover. He gave me no choice.'

Martyn was worried about his dad. He seemed to be working harder and harder, often to cover for unreliable members of staff. There had been a series of people who'd come and gone in recent months and the recruitment agency Peter had used to supply labour clearly didn't check out the staff they were sending to them. Most had moved from job to job and were available at very short notice. This helped the farm but often led Martyn to wonder why they were so readily available. George was a good example. Only 19, he loved his social life, but this didn't really mix with the need to be up early to get in the cows at 4 am.

'You should've told him to get a taxi, Dad. He's taking us for a ride. It's time we recruited our own staff anyway. These agency workers are a waste of time and money. The only one who's been any good is Marek and if we could find more like him, we wouldn't be chasing around after people like George.'

Marek was from Gdansk in Poland but had lived and worked in the UK for seven years. He'd work for six months, with just the occasional day off, and return home for a month. He worked hard, was excellent with machinery but also loved cows. He was a calming influence around the parlour and knew every cow's number as well as her characteristics. His English was proficient enough for him to get by and he was able to communicate effectively with the rest of the team, who all appreciated his attention to detail with the livestock. He'd even help Andrew during lambing season and never seemed to tire of the farm.

Martyn sighed. 'Dad, why don't you let Charlotte and me change things around a bit on the farm? There's so much we could do to help you and make your life easier. Mum's really worried about you too. You're not 25 anymore, Dad. You need to slow down and not be getting up early to move and feed the cows.'

Peter turned to head out of the kitchen, frustrated with the repetitive discussions that seemed to circle around the same narrative. 'Let's not start all this again, Martyn. The system me and your grandad set up on this farm works. It's kept me and this family in a safe and secure position for a long time. We don't need to change anything at all. We just need to work hard, and it will all come good, you'll see.'

But inside, Peter could feel a nervousness he'd not experienced before. He also noticed a tightness in his chest, which he'd experienced a few times recently. He dismissed both sensations as nothing more than indigestion and moved out of the kitchen towards the utility room. The constant smell of slurry and silage filled the air and Peter wondered if he'd ever enjoyed a day when his nostrils weren't full of the smell of the farm. He pulled on his overalls and wellingtons and headed out towards the parlour to see what the team were up to. He met Charlotte running towards him and waving her arms.

'The cows are out, Dad!' she shouted across the yard, pointing towards the fields beyond the parlour. By the time she'd reached Peter, she was out of breath and sweating profusely. She seemed agitated and ready to blow at any moment.

Charlotte was furious. 'It's ridiculous. I've told Amy a thousand times to ensure the gate at the bottom of the hill is pulled up correctly. But now the cows are trampling all over Mrs Prince's front garden, which will mean another claim against us. I think we should deduct the cost of this from her wages. I know she's helping me with the calves but all she needed to do this morning was help Marek bring the cows in. Can you give us a hand to quickly round up the cows?'

Peter agreed and they headed towards the Kubota utility vehicle parked in the yard just across from the house. On hearing the commotion, Martyn ran towards them, jumping in alongside his father. Charlotte climbed into the rear space, and they sped their way towards the far field where the cows had escaped after being milked, clearly attracted by the fresh grass in Mrs Prince's large garden. As they bumped along the route across the field, there was an awkward silence, which frustrated Charlotte even more.

'What's up with you two?' she asked. 'Everyone has noticed the atmosphere between you and, to be honest, I don't need it this morning. It's hard enough managing the staff without you two being frosty too.'

'Take it easy, sis. We're fine,' said Martyn unconvincingly. Peter didn't respond.

When they arrived at the field gate where the cows had escaped, Marek and Amy were already in the road, trying to encourage about 20 cows to return to the field rather than head towards the cows further down the lane. Andrew was on the other side of the cows with Bertie, his sheepdog, also trying to help round them up in the same way he would the sheep. Cows were less likely to move and could get far more jumpy than the sheep but Bertie was still running backwards and forwards diligently to Andrew's whistles and commands. After lengthy hand waving, gentle coaxing and coordinated encouragement of the cows, they were all safely returned to the field and the gate was closed.

'Well done, everyone,' said Peter. 'Thanks for reacting so quickly – it shows what can be done when we all pull together.'

Charlotte still looked angry and was determined to make her point. 'Can we all please check gates when we're moving cows?' She turned towards Amy. 'I reminded you about this the other day. This has been a total waste of our time and we had to leave Billy by himself in the parlour. I'm sure we all could've been doing better things, let alone the cost of putting Mrs Prince's garden right… again! It's just not good enough, Amy.'

Amy looked sheepish while Andrew frowned at Charlotte's abrupt approach. Martyn could see Amy was feeling uncomfortable and he didn't like awkward atmospheres, preferring to use humour to lighten the mood.

'No real harm done though, eh? Gave us all a little exercise, didn't it? I'm ready for the day now!'

'That's not the point, Martyn,' said Charlotte. 'This has happened before. Just remember, everyone – it costs time and money.'

Charlotte started to walk back to the Kubota and Martyn winked reassuringly at Amy, who by now was red faced and clearly embarrassed her mistake had been so publicly pointed out by Charlotte. Peter and Martyn clambered back into the Kubota, where Charlotte was waiting for them.

'Dad, we need better staff. These idiots are going to ruin our farm. I want us to sack Amy tomorrow and send out a warning sign to the rest of the staff that if they don't buck up their ideas, they will be out too.'

'Woah, slow down Charlotte,' said Peter. 'You can't just fire people when they've made mistakes. Let's look at ourselves first to see what we could have done differently. Was Amy trained well enough? Plus, there's employment law stuff to consider too.'

'Employment law? I really don't care about that. I just want Amy off the farm. She does my head in.'

Charlotte stormed off towards the parlour leaving Martyn and Peter standing side by side. Martyn broke the silence.

'Blimey, my sister has got a temper. Not sure where she gets it from, Dad, but she can't just go round losing her rag. It's unprofessional.'

Peter nodded, concerned that his daughter wasn't able to control her emotions. 'Yes, she's still got a lot to learn about managing people. But her heart is in the right place. She just wants what's best for the farm, Martyn.'

'So do I, Dad.'

Martyn walked back towards the parlour to see how Billy was getting on with washing down after milking. He wondered why he and his father couldn't see eye to eye on the future of the farm when they both wanted it to succeed. He wondered why, with so much in common, they weren't able to communicate this to each other. Martyn shook his head and his father noticed him doing so as he watched him walk away.

'If only we could communicate more effectively,' thought Peter. He turned back towards the house and decided to find Andrew, who was now back and bringing the ewes up to the lambing shed. Peter loved the lambing season, just like the calving season, when the farm was awash with new life and energy. While everyone was tired from the long hours, the young animals brought a freshness to the farm. Each animal's birth was different and each mother's approach to their newborn was different too. The lambs would stay with their mothers while the calves would be moved away soon after being born.

Peter felt a wave of emotion come over him as he thought of his son moving away from him. He realised the farm was surrounded by adults and their offspring, all faced with being separated from each other. And while this was the nature of farming, he wanted his own family to stay together. He was an only child and knew he would stay on the farm all his life. When he thought about his own family, he wondered if maybe he wasn't a good enough father and husband. His daughter was headstrong and too blunt while his son was too easily distracted and lacked focus. And Mary was just a born worrier who wanted everyone to get along without falling out. 'Should I have done more?' he wondered.

His mind wandered to what Mary had said. Perhaps he *was* a little stubborn and maybe Martyn did have a point about taking on new ideas. And Charlotte was right about Amy, as she'd been told numerous times to check the gates.

'All this family and people stuff is exhausting,' he thought. 'No wonder people look to outsiders to run their farms for them. They would have much less emotional attachment to these situations and they could more easily bring people together. Mary did say things were better on Anna and Joe's farm since they'd spoken to this chap they call the Chairman so perhaps Mary was right and it's time to find outside support. Maybe he'd be able to convince Martyn to stay.' Peter was lost in his thoughts when Martyn appeared again.

'Dad, I meant to tell you earlier: I've got an interview tomorrow so I'm not going to be able to milk in the morning. I've spoken to Amy and she's going to cover for me. It might give her a bit of space from Charlotte too. OK?'

'Sure, no problem,' said Peter. 'Where's your interview?'

'A big farm in Cornwall that's looking for a senior herdsperson. It's a step into becoming a herd manager and the farm is bigger than ours so there's a larger team too. I know one of the herdspeople as he's in a discussion group I belong to, so he's given me an overview of the farm and its issues. They are down with bovine tuberculosis too, so there's a fair amount of pressure on their milk yield, which has fallen since they lost their last batch of cows. I think I could make a difference, Dad, so I'm excited about it.'

Peter wanted to be excited for his son but inside his heart was aching. 'That's great, Martyn. Whereabouts in Cornwall is the farm?'

'Near St Austell. So obviously I'd have to move down there but they have a nice house to offer me. It's got two bedrooms and it's just off the farm. I'd have my own space rather than being under your feet every day!'

Martyn was trying to sound excited but inside his feelings were mixed. He didn't want to leave the family farm but couldn't see any other way of maintaining a positive relationship with his father. He felt cornered and with little choice but didn't want to show this to his stressed and exhausted father.

'OK, son, I hope it goes well. I'll check everything with Amy in the morning and we'll make sure George gets in too. You go and smash it tomorrow and show them what a great asset to their farm you could be.'

Peter placed his hand on his son's shoulder. 'I wish you weren't going but I understand why. This farm is just too small for you. My thinking is too small for you too. But I want you to be happy, so no compromises, OK? If the farm is a shambles, you don't go there, you hear me? You stay here and we'll work things out.'

'I think we've tried for a long time, Dad, so I think this is for the best thing for all of us.'

Peter sighed. 'OK, but there will always be a place for you here.'

'I know, Dad. And our farm will always be the first place I want to be. But let's see how this interview goes and we'll go from there.'

Martyn walked away, holding back tears again, and Peter watched his son take his first tentative steps away from the family farm. He recalled when he was a similar age but had known he would eventually run the farm. He'd always agreed with his father about how the farm should be run. They were both cautious in their approach to business, preferring to take decisions slowly and with all the research and detail they could muster. But Martyn was different. Peter yearned for a relationship with him like the one he'd enjoyed with his own father.

'Are you coming in for your breakfast, Peter?' called Mary from the farmhouse.

'Just coming.'

Peter pulled his gaze away from his departing son and walked into the house, removed his overalls and boots and entered the kitchen where Mary was busy preparing breakfast.

'I gather some cows escaped after milking,' said Mary as she moved some bacon around the pan.

'Yes, Amy forgot to check a gate on the far field and a few got out into Mrs Prince's garden again. Thankfully they haven't caused too much of a mess and most of them were still in the lane. We got them all back in but Charlotte was blunt with the staff again and with Amy in particular. Amy looked very uncomfortable about it all.'

Peter sat down at the kitchen table as Mary slid a plate of bacon, eggs, tomatoes and mushrooms in front of him. The smell was tantalising and Peter realised how much he enjoyed this part of the day. Mary always prepared breakfast to arrive when he came in from milking and there was a small window of time where they would chat about the day ahead and the plans for the farm.

'This looks lovely, Mary, as always.'

'Well, you make sure you eat it all up and don't skip lunch today. There's bread and cheese in the fridge and fresh ham too so if I'm not here, you make sure you come and have something to eat, OK? Anyway, how's Amy? She's a lovely girl – a bit forgetful at times but we can't afford to lose her. We're already short staffed, aren't we?'

Mary pushed her breakfast around her plate.

'I'm going to speak to the Chairman again today,' said Mary, who didn't look at Peter but waited for a reaction.

'Oh, are you?' said Peter. 'And what pearls of wisdom do you think this mysterious Chairman chap can offer you now?'

Peter's sarcastic tone frustrated Mary but she accepted he wasn't ready to listen yet.

'As we've already discussed, he suggested we take on a farm manager and he recommended a business specialising in farm recruitment and staff development. I'm going to speak to James Glover from REAL

Success and ask them to place an advert for us. James comes highly recommended by several of our farming friends and his business focuses purely on helping farmers find and develop their people. Hopefully, an advert for a farm manager could go live later.'

It was Peter's turn to push his food around his plate, and he looked up at Mary. 'So why do you need to speak to this Chairman chap again, then? And how much is all this going to cost? We're tight against our budgets as it is.'

'It costs nothing to call the Chairman – he doesn't charge for his advice. He's been successful and now enjoys helping other business owners. Anna said that REAL Success charge a small admin fee to take on an assignment but we only pay them a fee if they find someone we take on. They do all sorts of extra tests and screening to make sure the person is the right fit for a farm. She said this chap James is really helpful, so I've got a good feeling about this. I think getting someone in to run our farm could be the best decision we've ever made.'

'*We've* ever made? Sounds to me as if you've made the decision already, not me.'

'Well, if I waited around for you to make a decision, I'd be doing so after your funeral and I'm not going to let that happen, OK?'

Peter smiled at Mary. 'OK, love. I know you only want what's best.'

Mary felt relieved that Peter didn't push back on her decision to recruit help. But she hadn't told Peter everything about their first conversation. The Chairman had also said that without help for Peter, the farm could begin to suffer as he'd seen it happen time and time again. He'd suggested that when the farm owner holds on to the running of the business for too long, the farm suffers but the farmer suffers too. He said too many farmers have no life outside the farm and that some make themselves ill by not planning enough rest. Many never retire as they have nothing else but the farm in their lives.

Mary felt instinctively that Peter was under too much pressure and she would lie awake at night, feeling anxious about his health. Something needed to shift and for the first time in a long time, Mary felt the whisper of change around them. And it had all started with her first conversation with the Chairman.

4

Andrew loved his job. The lambing season was his favourite time of year and despite having been a shepherd for more than 20 years, he never tired of the sight and sound of lambs being born. He prided himself on losing hardly any newborns, despite a few occasionally dramatic births needing the help of the vet. Twins were often the biggest challenge with tangled legs or one being round the wrong way. But they rarely lost an animal and even the weakest were given the chance to grow.

However, lately Andrew had been feeling differently about his role and the farm. For years, everyone got along, happily doing their jobs and focusing on what needed to be done. But since Charlotte and Martyn became more involved, the atmosphere had changed. Charlotte was far too hot headed and rude at times and Andrew knew Arthur would be turning in his grave at the way she spoke to the staff. Arthur had always respected the staff, recognising the farm didn't work without them and treating everyone with respect. Peter had followed in his father's footsteps and until recently, Andrew had never considered anything other than retiring after only ever having worked at Wilson Farm.

With no family other than a brother who lived in the north-east of England, the farm *was* his life. Being out in the fields tending to his sheep was his happy place. Peter would leave him to do his job with little interference but Charlotte had started to meddle. What frustrated him the most was that Charlotte knew very little about sheep and how to get the best from the flock and yet she still insisted on expressing her opinions, often aggressively. Watching Charlotte shout at Amy in front of everyone else got him thinking for the first time in his adult life that he might want to work somewhere else.

He leaned on his trusty shepherd's crook, watching the ewes feed the newborn lambs in the shed. The calls of young lambs searching for their mothers and the mothers' response was a melody that filled his spirit.

But there was also a heaviness that pulled him down. He wasn't sure if it was mild depression or simply sadness brought on by the changes at the farm but compared to the vibrancy of life in front of him, he couldn't understand why he felt so low.

'I think I'll have a chat with Peter,' he thought. 'I need to tell him how I'm feeling. If she doesn't change her manner and approach to the staff, his daughter is in danger of harming the farm. And I'm not going to accept being spoken to the way she spoke to Amy.'

'Morning, Andrew. How are things going here?'

It was Charlotte. Andrew felt awkward as his thoughts had been so negative about her but he didn't want it to show.

'Things are going well thanks, Charlotte. Six more born overnight and mums and babies are all doing well.'

'Have the lambs all fed well?'

Andrew felt a surge of anger and frustration building within him. 'How long does she think I've been doing this job?' he wondered.

'Yes, Charlotte. All the lambs are feeding well and I've checked them. I've been doing this job for a little while now, so you don't need to worry yourself with the lambing. I've got it covered.'

Andrew was barely able to mask his irritation with Charlotte's patronising tone but he hoped a weak smile as he finished his sentence would suffice.

'I know you've been our shepherd for a long time, Andrew, but there are always things we can do better. I was watching a YouTube video and the vet was saying how important it was for the lambs to get their colostrum within a few hours of their birth. I was just checking, that's all.'

'YouTube!' thought Andrew. 'People think they can watch a video and become experts overnight.'

'OK, Charlotte, but I'm all good here thanks. No need for you to be checking up on me.'

Andrew was struggling to keep himself in check.

'Can I ask you a question, Charlotte?'

'Of course.'

'How do you feel about the team you've got here at Wilson Farm?'

Charlotte looked perplexed. 'How do I feel about them? Not sure what you mean, Andrew. My job isn't to feel anything about the staff. It's to make sure they do their jobs correctly and most of the time they don't.

So if that's what you mean then I feel we haven't got the right staff to do the job in the way I want it done.'

Andrew sighed inwardly at Charlotte's cold approach but it told him all he needed to know. She simply didn't recognise the staff as having any value. He thought about Arthur and how sad he would be that his granddaughter appeared so heartless and uncaring.

'But don't you think your staff are really important to the farm, Charlotte?'

'The only thing important to me is producing milk to go in the tank, lambs we can sell and crops to feed our livestock,' she replied. 'That's got to be our focus, Andrew. It's certainly my focus. And if the team were to follow the protocols and instructions I've laid out rather than thinking they know better then, to use your words, I'd *feel* far happier.'

'OK, Charlotte. I hear you.'

But Andrew didn't 'hear' Charlotte at all. All he could hear in his head was the sound of staff resigning and, in his mind, he saw Peter's face growing pale with concern and more stress.

'Thank you, Andrew. I'm glad we're on the same page.'

'Gosh,' he thought. 'We're about as far away from being on the same page as it's possible to be.'

Charlotte walked away with her usual air of defiance and sense of purpose. Andrew wondered where her direct manner came from as neither Peter nor Mary were like that. He was just about to find Peter to express his concerns about Charlotte and the effect she was having on the farm when Peter entered the lambing shed.

'Morning, Andrew, how are things this morning? Everything looks very calm and under control to me.'

Andrew was struck by the contrast in the two family members' approaches. One was keen to challenge and criticise while the other was keen to praise and encourage.

'It's going well thanks, Peter. Six more lambs this morning and mums and babies are all going strong. We've not had a single rejection yet so I'm very happy with the season so far.'

Andrew decided this was the moment. 'Peter, can I talk to you about something that's been bothering me?'

'Sure, Andrew, of course.'

'It's about Charlotte.'

Peter felt his chest tighten a little and he braced himself for what Andrew might say.

'What about Charlotte?'

'As you know, Peter, I've worked here all my life and I think of you as family, as much as any of my own. And I don't want to speak out of turn or cause a fuss but I'm worried that Charlotte is upsetting the staff so much that they might leave.'

Peter knew that Andrew was right and felt his chest tighten a little more. 'How about you, Andrew? Have you considered leaving?'

There was a pause. Andrew wondered whether he should deliver this blow to an already exhausted farmer whom he'd watched grow under the stewardship of his father into the steady and successful farmer he was today.

'This farm is my life, Peter, and I can't imagine ever working anywhere else, but this morning, for the first time, I did wonder whether being a shepherd elsewhere might be something for me to consider. Nothing concrete or decided, you know, but just a thought that came into my head.'

Andrew watched the colour drain from Peter's face. It was clearly a hurtful blow and he was already regretting what he'd said. Then he saw Peter reach out and grab hold of one of the lambing pens.

'Are you all right, Peter? You look very pale. I'm so sorry if I've upset you.'

Peter wasn't listening to Andrew. He grabbed onto the pen with his other hand and then fell to his knees. The world was starting to spin around him and the tightness in his chest had become a pain in his left arm. He felt lightheaded and could sense the cold sweat running down the back of his neck.

Andrew tried to grab Peter as he started to collapse but he was soon on the ground holding Peter in his arms.

'Peter, Peter! Are you all right? What's wrong?'

'Get Mary,' was all Peter could whisper. 'Get Mary, quick…'

Peter could feel his eyes become heavy and he couldn't stop them closing. He felt calm but tired.

'Peter! Peter!' shouted Andrew.

But Andrew's voice was distant now. The sounds of the shed were diminishing too as the sweet melody of the lambs became a lullaby. Peter felt a warm glow of peace surrounding him as he gently closed his eyes and drifted into a deep and welcome sleep.

5

As he opened his eyes, Peter became aware of a faint sound of buzzing and bleeping. He had no idea where he was. Then the room started to come into focus and he could see a light above his head.

'Dad, Dad, can you hear me? I think he's waking up, Mum. Look, his eyes are flickering.'

Mary moved closer to the bed and touched Peter's hand.

'Welcome back, love. You're in the Derriford. You've had a heart attack. Can you hear me?'

Peter felt as if he was travelling down a long farm track with his family calling him from a distance. As his vision stabilised, he could see his beloved wife standing over him. He squeezed her hand.

'You've had a narrow escape, love,' she said. 'Just a mild heart attack but it's knocked you off your feet and you've ended up here. It was a good job Andrew was with you when it happened.'

Peter's mind was fuzzy. He remembered speaking to Andrew about the lambing but then everything went blank. There was something Andrew wanted to tell him but he couldn't recall what it was.

'Hey, Dad, you gave us all a fright,' said Charlotte, trying to make light of the dark situation she felt partly responsible for creating. 'Although the paramedics were wonderful, Dad, so you were well looked after!'

Peter looked around and saw that his two children were also there with him. He felt loved and cared for but was confused as to exactly what had happened.

'Mr Wilson, my name is Vinay Parmar. I'm the registrar here on this ward.'

Peter turned his head slowly and saw a young doctor standing next to his bed.

'You're in Ward 11 at Derriford Hospital. You've had a mild heart attack, which is a warning to you, Mr Wilson. Mrs Wilson here tells me you've been working too hard and your body has just confirmed that to you. We're going to run a few tests on your heart to make sure all the valves are working as they should be and your arteries aren't furred but I'd say you had a lucky escape this time.'

Peter thought the doctor looked young enough to be in short trousers but he seemed to know what he was talking about. As his mind began to clear he wondered why he'd suddenly experienced a heart attack.

'What's caused it, doc? I'm a fit man, aren't I? I work outdoors and I'm active so why did my ticker decide to object?'

'It's very simple, Mr Wilson. Stress produces hormones and chemicals including adrenaline and cortisol, which can cause damage to your heart over a long period of time. If you don't reduce the stress, you keep producing these chemicals. After a while, your body can't cope and something blows. It could be your heart or other organs that begin to fail. Irritable bowel syndrome, gut issues, weight gain and lots of other conditions can be caused by the build-up of stress in your system. You were lucky this time but you might not be so lucky if it happened again. Lifestyle changes are in order for you, sir.'

'Yes, Dad,' said Martyn. 'We've been telling you to ease up. I've been trying to tell you about ways to improve the farm to make things easier but you wouldn't listen to me. And now look at where you are.'

'Now's not the time for blame or "I told you so",' said Mary. 'We just need to get Dad home and then we can discuss how we might run things differently to give him more time and less stress.'

'Yes, Mum, but he won't listen,' said Martyn. 'I've been telling him for ages that we need to make changes on the farm.'

'Stop it, Martyn,' interjected Charlotte, who'd been silently observing her father for a few minutes. 'Dad needs peace and quiet, not more angst around him. Anyway, Mum and I have been talking and we agree that we need extra help around the farm. I was talking to Andrew this morning and he gave me feedback about the importance of having the right staff. And I realised I'm not ready to run the farm yet. You're looking for a new job too so we will need to recruit a farm manager. And Dad, no arguments, OK?'

Peter was still waking up and felt too tired to argue with his family. He just wanted to rest and not make any big decisions. However, despite his lack of energy, he sensed something needed to change and, finally, wondered if the family could be right.

'OK you lot. You win. I'll leave you to sort out the details. Now just let me rest and get some sleep.'

Peter put his head back on the pillow and closed his eyes while his family began to speak in hushed whispers. Martyn spoke first.

'I can't believe he finally agreed we should get him help, Mum. That's awesome. Have you contacted James Glover yet?'

'Do we really need a recruitment consultant, Mum?' asked Charlotte. 'Can't we just ask around locally? It could be quicker and cheaper.'

Mary had contacted James the previous day and set the wheels in motion. Hopefully, an advert was already live and attracting candidates. She was convinced that finding external help was a good idea.

'I think it's right to use a specialist with expertise. If an animal gets sick, we call the vet. When we need business advice, we call the accountant or the bank manager. When we need technical advice around the farm, your father calls John Hemmings. But we have no one to call when things go wrong with our staff, so I think a people consultant is a good idea.'

As they sat around Peter's bed, they discussed the sort of person who would be suitable for the farm.

'I think we need someone decisive who will kick ass,' said Charlotte. 'There's no point in just trying to please everyone, is there?'

'No, that will just upset everyone,' said Martyn. 'We need them to be open to new ideas, friendly and ready to listen to others, rather than trying to boss everyone around.'

'You might both be right,' said Mary. 'Maybe they need to be able to make decisions for the farm after carefully considering all the options. Someone who is good with people and open to new ideas?'

'Blimey, it sounds like we each think we need someone different. Does this sort of person really exist, Mum?' asked Martyn.

'I doubt it,' said Charlotte.

'I'll speak to James and see what he says,' said Mary. 'He mentioned a test they use to help them understand the character of the people they're suggesting we consider. He also asked us all to take the test so he knows more about us too.'

Mary didn't mention to Charlotte and Martyn that in her conversation with the Chairman, he also mentioned the importance of finding the 'right' person for the role on the farm and not being tempted to take ‘anybody’. He’d said this was where most farms came unstuck as they would recruit quickly because they were desperate to fill a role and then discovered the person either couldn’t do what they said they could do or simply didn’t fit into the team. The Chairman referred to it as ‘recruiting in haste and regretting at leisure’.

They continued to discuss how a farm manager would help the entire team work together more effectively and assumed Peter had been sleeping throughout their hushed conversation. But Peter had heard every word. He had kept his eyes closed but was fascinated by their different perspectives on the type of person who would best fit the farm. His exhausted body and mind simply made him realise he was tired of making all the decisions, tired of having to research everything and ensure the data was correct, tired of managing relationships on the farm and tired of having to consider new ideas. Whoever was going to join Wilson Farm needed to be a special person. Despite being sceptical that such a person existed, his physical and emotional state told him that it was time to let go of his resistance to getting help.

As Mary, Charlotte and Martyn left the hospital to head back to the farm, having gently kissed their husband and father goodbye, they continued to debate the sort of person who would best suit the farm. Without coming to any conclusion, they all agreed it was best to wait until James Glover proposed his first candidate.

It had been a tumultuous time for the Wilsons. But as Mary, Charlotte and Martyn travelled home, each of them felt that something had shifted. Like gears changing in a clock that’s about to chime, life at Wilson Farm was about to change. The family was, for the first time in a long while, united.

6

Billy, George, Andrew, Amy and Marek were having their morning break together. Andrew told them about Peter's heart attack and how they all needed to rally around the family and pull their weight. Andrew was the longest-serving member of the team after Billy and old enough to be the father or grandfather of most of the staff, so when he spoke, they listened to him. His calm demeanour and gentle style ensured that no one found him difficult to get along with.

'It's easy for us to forget just how much pressure Peter and Mary are under,' said Andrew as they all sat drinking tea in the small staff room. There were graphs and statistics printed on A4 sheets of paper and stuck on the walls with Blu Tack along with the rota for the next two weeks. 'These numbers are only half the story,' he said, pointing at the charts. 'In addition to our own efforts, a whole lot of work goes on to make these results happen.'

'I know I've been a bit slack recently,' said George, looking down at the floor. 'I'll make sure I'm on time now. Sorry everyone. I'm a bit of an idiot at times so I apologise if I've let you down.'

George was keen to make a good impression but also to enjoy his social life. He lived at home with his family just five minutes away from the farm. His father had left the family home just over a year ago to settle with another family and this had hit George hard. He idolised his father so when his dad walked out, his world crumbled. He also needed to support his mum and brother as he was the eldest. He'd found solace in drinking at the local pub with his mates but this often turned into late night binges, resulting in him not making it onto the farm the next day.

'We know it's been hard for you,' said Billy. 'It's cool. Just try to make sure you're here when we need you. That's all we ask.'

'Working hard is good for the family,' said Marek. 'If we all do our jobs well and the farm runs smoothly, no problem.'

‘I screwed up again this week with the gate,’ said Amy sheepishly. ‘I hope it didn’t push Peter over the edge, because Charlotte was proper angry, wasn’t she?’

‘It’s OK, Amy,’ said Andrew. ‘We all make mistakes but the key is to learn from them.’

The team continued to chat about the day ahead, the tasks that needed to be completed on the farm and some tweaks to the rota. George agreed to help with feeding the cows and Amy agreed to do a couple of extra milkings. Andrew was working flat out with the lambing and Billy agreed to help him for the next few days. Marek was happy to stay in the parlour and focus on the cows. Between them, they managed to cover most of the jobs without Martyn and Charlotte.

Just as they were all about to get up and continue their day, Mary entered the staff room.

‘Morning everyone. I just wanted to check everyone is OK after our shock yesterday?’

‘We’re all fine,’ said Andrew. ‘But more importantly, how’s Peter? And how are you, Charlotte and Martyn coping with the shock? I really thought he was a goner, you know.’

‘We have a lot to be grateful for,’ said Mary. ‘I’ve no idea what might have happened if he’d been on his own somewhere around the farm. He might not have made it.’

Andrew thought Mary looked tired but remarkably calm given that 24 hours earlier the farm had been full of paramedics and upset staff. By the time the ambulance arrived, Peter was drifting in and out of consciousness and his lips were turning blue. Mary was in shock, Charlotte was too upset to help and Martyn simply stood motionless as if watching a movie play out in front of him. Thankfully, Andrew had remained calm. He and Marek had put Peter into the recovery position so that by the time the ambulance crew arrived, they were able to quickly take over.

‘Thank you all for everything you did yesterday,’ said Mary. ‘I hope it wasn’t too shocking or upsetting. Do come and talk to me if you have any questions. Peter should be back home in a few days, so in the meantime, we all need to pull together.’

Andrew raised his hand. ‘We’ve talked together this morning and sorted out the rotas. We can cover everything, even if Charlotte and Martyn don’t feel like working.’

Mary's heart filled with pride. Andrew was such a loyal employee and had become like an older brother to Peter and Mary. Because of his calm manner and organised approach, they left him to run their sheep enterprise without any interference. It was a relationship based on mutual respect and trust. In the current circumstances, she knew Andrew would step up and that was exactly what he'd done.

'Thank you, Andrew,' said Mary. 'Thank you, all of you. I have some news to share with you too.'

The group shuffled in their seats as if they were anticipating bad news. Amy looked down, expecting to be told she was going to lose her job. George looked out of the window feeling guilty that he'd let the family down a few times recently. Billy tickled Bertie's neck while Marek sipped his tea and Andrew simply raised his eyebrows.

Mary continued. 'As a family, we've been trying to convince Peter to take on a farm manager. Since Grandpa Arthur died, Peter has been covering his work and Martyn has also decided he wants to work on another farm for a while. He's been finding it hard to get his dad to accept his new ideas and they've been arguing so much that they've decided between them it's best if Martyn gains experience elsewhere for a while. All of this is leaving us a bit short on labour and experience so if Peter is going to slow down, we need someone who can come in and take the pressure off him.'

Mary paused to assess the reaction from the team but there was just stunned silence. Mary broke the ice. 'So, what do you think?'

Amy spoke first. 'I'm sorry I forgot to close the gate yesterday. I hope it didn't add to Peter's stress. I feel awful. I hope it wasn't the cause of his heart attack.'

'No, of course it wasn't,' said Mary. 'And there's no point going over it again, Amy. What's done is done. But what do you think about a new farm manager idea?'

'I think it makes perfect sense,' said Andrew. 'We've needed some help for a while. I could see it wasn't going to work with Martyn and Peter. With the greatest respect to Charlotte, she has a fair bit to learn and you're already fully employed in your role on the farm as well as looking after the family. It's a good idea, I say.'

'Thanks, Andrew,' said Mary. 'Anyone else?'

'I'm easy,' said George. 'Whatever is best for the farm is fine by me.'

'And you, Billy?'

'Of course I'm fine with it. Andrew and I have said many times that you need more help. Just make sure it's not a bright-eyed youngster who think they know everything but have lots to learn. We know how things work around here. We don't want someone coming in throwing their weight around.'

'I understand, Billy. I really do. I don't want that either. And you, Amy?'

'I'm just so sorry about what happened. If the new person can come in to help us all, I'm sure it will be good for everyone. What does Charlotte think?'

Mary was aware that Amy and Charlotte did not see eye to eye. Or rather, Charlotte didn't respect Amy, so she wasn't surprised Amy was keen to hear Charlotte's opinion.

'Charlotte still has a lot to learn, Amy – particularly around getting the best from the team. She knows this and wants what's best for the farm and her dad. So, she's behind the idea. Marek?'

'Peter works far too hard. He works harder than us Polish workers and that is hard! So I hope this man can come in and make life easier for Peter.'

'It could be a woman, of course,' said Mary. 'It doesn't have to be a man. I've been in contact with a couple of people Anna from Littlebeth Farm recommended and one of them, James Glover from REAL Success, is going to help us find the right person. To be honest, it has been ages since I interviewed anyone, so I hope Peter is back on the farm by the time we need to interview them!'

'You'll be fine,' said Andrew. 'We can all help too. I'm sure the new person would want to meet us anyway. We'll all be on our best behaviour, don't you worry!'

'Thanks everyone. I really appreciate your support. I'll let you get on with the rest of the day now. See you later.'

Mary exited the staff room and there was an audible exhalation from the team.

'Phew,' said George. 'I thought she was going to fire one of us. I don't have an issue with a new farm manager.'

'Yeah,' said Amy. 'I thought I was going to lose my job today so I'm relieved, to be honest.'

The group continued to discuss the implications of having a new farm manager and the opportunities they could bring to the farm. They exchanged views on the type of person they wanted and were surprised that their wishes were slightly different.

Andrew said he hoped they would be organised, structured, could work to plan and would let him get on with his job without too much interference. Billy agreed with Andrew but added that he hoped they would also have a friendly and calm manner. Marek wanted someone who would put together a daily plan and George didn't really care who the person was as long as they weren't too serious and allowed flexibility around the rotas. Finally, Amy hoped for somebody who was calm, friendly and asked her to do jobs rather than barking orders like Charlotte did.

Despite the worry of the previous 24 hours, they all left the staff room with a spring in their step. There was an air of expectancy and, for the first time in a while, the team was united in spirit and commitment to a cause.

7

Mary spent most of the morning making phone calls. She needed to tell their mutual friends about Peter's heart attack as she knew they would want to know. She also rang Abigail the vet, Jim Stanley, the accountant, and John Hemmings, who all expressed deep concern for Peter and offered their support.

Mary also called James Glover, who was upbeat about the advert he'd placed online along with enquiries he'd made about potential candidates. While it was early days, he felt confident he'd be able to find the right person to fit the team. He'd asked Mary if everyone could take a personality test to help him match candidates to the team. He offered to visit the farm to run a short session where they would all discover their personality styles and how this affects the way they work together. He explained that people find that they can work more easily with some colleagues than others and the session would help everyone work together more effectively as well as help James understand the type of character who would be best suited to be their farm manager.

'The Team Dynamics session is one of our most popular,' said James. 'Even the most diehard sceptics in the team will enjoy it.'

Mary had reservations about how the team would respond to being analysed and scrutinised but she agreed to book a session the following week. They finalised arrangements and James said he'd keep Mary up to date with progress.

When the call with James concluded, Mary decided to contact the Chairman again. She pulled out her notepad where she'd made notes from her previous conversation. She'd written down two phrases to remind her of what the Chairman had said.

Find the silence.
Recruit in haste – regret at your leisure.

Mary wondered what the Chairman would say to her this time. After a couple of rings, he answered in his warm voice.

'Morning, Mary. Good to hear from you again.'

Mary was a little startled that he recognised her voice and guessed her number must have come up on his phone.

'I'm good, thanks,' Mary responded. 'Before we go any further, can I ask for your name? It seems very impersonal to know you only as the Chairman.'

There was a short pause, which felt like a lifetime to Mary. Had she upset him? Did he feel it was inappropriate? Her thoughts tumbled around her head in the few seconds of silence that followed.

'Of course, Mary. My name is David. Most people just call me the Chairman as this is how I was known in my last role but you can call me David if you'd prefer.'

Mary breathed a sigh of relief as their relationship now felt more personal.

'OK, David. I've spent time thinking, or "in the silence" as you put it, since our last call and I think I've come up with a good decision.'

'You think you have or you know you have? There can only be one or the other. If you're uncertain, then the decision either isn't fully made or it isn't the right one. If you know you've made the right decision then your gut, your instinct or your intuition, however you wish to describe it, will tell you so. What does your instinct say about this decision you've made?'

Mary remembered a woman called Sarah who'd applied for a role on the farm several years earlier. She was 10 minutes late for the interview and said she could milk cows, trim feet and had passed an artificial insemination course. At the time, Mary felt instinctively that she wasn't telling the truth but Peter was desperate to fill the role and Gramps encouraged them to take her on. It proved to be a disaster. After just three months, they had to let her go.

'It feels right,' continued Mary. 'I've gone over it in my head many times and sense that it's the right thing for us to do.'

'When dealing with people, your intuition is one of your most powerful tools,' said David. 'Learn to trust this more than your mind as your mind is simply looking for solutions to problems. It wants to keep you safe but your intuition wants what's right for you, which may not always be what appears at first.'

Again, a memory stirred within Mary. She remembered how she'd felt about George when he came for his interview. He was young, with an active social life, living locally rather than on the farm, and yet there was something about him that she found interesting. He'd endured a tough upbringing but she sensed he had a good heart, which was important when looking after livestock. She'd encouraged Peter and Charlotte to take him on, even when they both said he was a bit 'over the top' with his enthusiasm at the interview. While there had been a few ups and downs with his timekeeping and reliability, Mary could see something in George that encouraged her. It was his hunger for knowledge and a willingness to step outside his comfort zone and try new things.

'Thanks, I will bear all this in mind. Oh, sorry – I mean, I will trust my intuition.'

'Well done,' said David. 'Has James Glover supplied any CVs for you yet?'

'I'm hoping to see the first batch later today.'

'Remember to look beyond the words on the CV. How does the person express themselves? Is there a passion within them to learn? What are their interests outside of farming? There's always much more to a person than they can put onto a CV, so read between the lines and trust your intuition.'

'I will. Thanks for listening and for your advice, David. I mean, Chairman.'

There was a gentle chuckle at the other end of the line.

'Go well, Mary,' he replied. And then he was gone.

Mary stood holding her phone in her hand for a few seconds before looking down to check that the call was over. She wondered about David/the Chairman. There was certainly a presence to his voice and a resonance that caused you to sit up and listen. His advice so far was more… what was it? 'Deep,' answered the voice inside her head.

She was glad Anna had suggested calling the Chairman. While his advice so far hadn't been purely practical, it did make sense to her. He was bringing another dimension to her thinking alongside the practical work James Glover was doing. It was an interesting blend of practical and emotional or even, yes, 'deeper' support.

It had been a busy morning but despite the tension and stress of the previous day, Mary somehow felt lighter – even excited. It struck her that during one of the most frightening times of her life, something new and magical seemed to be happening.

'A fascinating paradox,' she thought to herself and turned towards the kettle to make herself a well-earned cup of tea. She decided to add a couple of notes from her conversation with David. Now her notes read:

Find the silence.
Recruit in haste – regret at your leisure.
Read between the lines.
Trust your intuition.

As the kettle boiled and Mary sat in silence at the kitchen table, she was aware of a shift in her awareness. She pictured two train tracks running in parallel and it felt as if she'd just switched tracks onto a new set to take her towards a different destination. It was a comforting image of gentle but important change.

8

Peter woke up and, after a few moments, realised he was still in hospital.

'Good afternoon, Peter. How are you feeling?'

A young nurse was standing at the end of his bed reading his notes. She looked younger than Charlotte, with her brown hair scraped back into a bun. She wore no make-up but offered him a big smile, which lifted his spirits.

'I'm good, thanks. Feeling better after a good sleep. To be honest, I've not been sleeping well recently with everything that's been happening on the farm.'

'Well, it sounds like that's just what you needed then. I'm Grace, by the way, and I'll be looking after you today. I just need to do a few observations and then I can let you get back to resting, OK?'

Peter was happy to let Grace take his temperature and blood pressure as her cheery demeanour was a pleasant distraction from what needed to be dealt with back at the farm. Through the clearing haze of sleep, he started to think about Martyn leaving and Mary's suggestion that they needed to appoint a farm manager. Once Grace had left him alone, his mind started to run through various scenarios including Charlotte leaving the farm after her brother if she couldn't work with the new farm manager. Then the family would start to crumble, Mary would decide to divorce him and his life would fall apart.

'Morning, Mr Wilson. My name is Benjamin and I'm the hospital chaplain. I just wanted to check in with you after your health scare and see if you have any questions I could answer for you?'

Benjamin was a tall, handsome black man with very dark skin and a strong accent that Peter suspected was of South African origin. He was completely bald and Peter estimated he was around 40 years old.

'I'm good, thanks, Benjamin. Looking forward to getting home as soon as possible to be honest.'

'There's no rush though, is there? You can get a little rest here in this ward; it's exactly what the patients need most. Do you have a family, Mr Wilson?'

'I'm Peter – no need to be so formal, Benjamin – and yes, I have a wife and two children.'

'You're a lucky man, then. Family is a blessing from God we need to treasure, Peter.'

Peter wasn't a religious man and while he loved his family, he didn't see them as a gift from God. His family were his inspiration at times and his frustration at others. Martyn had been quite a burden over the past few months and Peter couldn't deny he was quietly pleased that he was, hopefully, going to be working on another farm for a while.

'I am indeed a lucky man, Benjamin. I nearly checked out yesterday so still being here is a big bonus.'

Benjamin held a book in his hand, which Peter assumed was a Bible, and he noticed a small cross around his neck. He wore a simple linen suit and while everything about Benjamin was downplayed, he exuded an energy of calm reassurance that Peter found comforting.

'Whatever your own belief system, Peter, I believe God has a plan for all of us. This heart attack was part of His plan. What do you think God might have been trying to say to you?'

Peter didn't need God to be saying anything to him. He knew he'd been working too hard and the family had now convinced him to take on a farm manager, so he didn't think he needed any divine intervention and certainly didn't believe his heart attack was divinely inspired.

'I don't believe in God I'm afraid, Benjamin, but I respect that you do. However, my heart attack was simply caused by working too hard and not looking after myself. I'll be fine in a few days, you'll see.'

Benjamin scratched his chin. 'Can I tell you a story, Peter?'

'If you must,' thought Peter, worried that Benjamin was going to give him a lecture on Christian doctrine. 'Go ahead,' he said.

'When I was a child in Mozambique, my father worked very hard on our farm. You see, my family are farmers too. But after independence in 1975, our country descended into civil war and many people were forced to leave. My family stayed but things became very bad. My father worked hard on the farm to look after us, just like you have. But eventually, he became very ill and passed away. He believed in God and we all went to church. But when he died, I turned away from God and my faith. My father was a strong man who loved his family and his country. But so much changed after independence; I think he became tired with it all. When he died, we moved to France and then to the UK, where our family has been for the past 15 years. I found my wife here in England and we now have two children – Charles and Anne. We named them after the Queen's first two children.'

Peter was beginning to wonder why the chaplain was telling him his life story and struggled to stifle a yawn.

'I'm telling you this story as family is the most important gift we have. I lost my father because he worked too hard. Our country was a bad place to live and we needed to leave and build a new life. But if I hadn't left Mozambique, I would never have met my Sarah, I would never have found my faith again, and I wouldn't have my children. So, you see, sometimes there's a bigger plan in what happens to us. I believe you must see the bigger plan for you in your heart attack. From it, good things may come.'

'I'm so sorry to hear about your family and what happened to your father,' said Peter.

'It's OK. I feel settled with it now as my life makes sense – but often our life only makes sense backwards. At the time we experience challenges in life, we can see them as barriers to our progress or emotional traumas to punish us. But in hindsight, when we look back over our lives, often the most challenging times are the times when we grow or stretch ourselves the most. It's not possible to see this at the time but this is what I mean by God's plan for us. It can make no sense to us but later, when we look back, it does. Anyway, I've taken too much of your time this morning. I'm glad you're feeling rested and send God's blessings to you and your family. Remember, Peter, families go through ups and downs – it can break us or grow us. Choose this time as a time of growth for yourself, Peter – not just in your business but in your life, too.'

Peter realised he'd judged the man who'd sat down next to his bed and a feeling of guilt engulfed him. Benjamin was clearly trying to help him and yet his Christian perspective was something Peter feared rather than invited. But the chaplain spoke such sense in the short time he'd spent with him, Peter knew it was a significant moment.

'Thanks, Benjamin. I really appreciate you coming round to say hello. It's been a pleasure meeting you and thank you for sharing your family's story with me. It's so easy to imagine your own pressures in life are unique to you and it's always helpful to hear of others' experiences. It has given me a greater sense of perspective. Thank you once again.'

The two men shook hands and Benjamin moved across the ward to speak to another patient. Peter watched him as he gently moved from bed to bed, no doubt telling a different story or passing on a different message to everyone he met.

Peter thought about his own family and his heart attack. 'Perhaps it was a nudge rather than a warning?' thought Peter. 'Maybe the heart attack was to prevent something more serious happening?' Peter wasn't sure but Benjamin's story of his life in Mozambique and his father working too hard on his farm most definitely struck a chord.

'Good afternoon, lovely husband!'

It was Mary, carrying a bag containing fresh clothes and some fruit. 'How are you feeling today?'

'Good, thanks. Just enjoyed an interesting chat with the hospital chaplain, Benjamin.'

Peter pointed to where Benjamin was sitting, on a chair next to an older man whose eyes were closed. Benjamin was holding his hand and Peter wondered if he was praying for him.

Mary looked surprised. 'I didn't think you'd be interested in talking to a chaplain, Peter. You've not stepped foot inside a church for 30 years!'

'I know, Mary, but he told me about his father, who ran a farm and worked so hard he eventually died. Then he told me about leaving his home country and ending up in the UK, where he met his wife and had his two children. He said his father's death was the catalyst for a series of events in his life. In other words, something amazing was initiated by something awful. He told me to look at my heart attack in this way.'

Mary unpacked the bag of clothes and placed the fruit on the small vanity unit next to the bed. 'Sounds like a very sensible way to look at it. And, in fact, it's how I felt after my last conversation with David. The world is full of paradoxes. What we see as the worst times in our lives can be the birth of something new.'

'Who's David?' said Peter.

'Oh, he's the chap I was telling you about,' said Mary. 'The Chairman.'

Peter remained sceptical about anyone who was referred to as 'the Chairman'.

'I thought you were speaking to a recruitment guru chap as well?'

'I'm not sure James Glover would see himself as a guru but yes, he's started to look for us. I'm hoping to see a few CVs later.'

Peter's heart filled with pride for his wife. She'd often been seen by many as a minor partner in the business. Whenever suppliers or contractors called and asked to speak to the farm owner, she accepted that they didn't consider her to be the person to speak to. Many expected to speak to the man on the farm and while this frustrated her, she duly went along with the expectation and put them through to Peter. But Peter could see she was taking control of farm affairs and making important decisions. She'd always done this but often in the background, helping Peter at key moments. But now she was in the foreground where she rightfully belonged and Peter was delighted and grateful for her optimism and decisiveness.

'Thanks, love. And how are things back on the farm?'

'The team have all stepped up. Andrew called a staff meeting this morning and they've worked out rotas to cover for you, Charlotte and Martyn. They all shared their concern for you and it's a real joy to see them pulling together. It's amazing how a crisis can do this, isn't it?'

'Yes, it is. We're so lucky to have people like Andrew and Billy who've worked with us for so long. They know the farm as well as we do.'

'And I've told them about the recruitment of the farm manager. They were all supportive of this too. So, it's all systems go at Wilson Farm. And you're not even there!'

Mary and Peter laughed together and chatted for a while about farm-related issues but also about the fact that they hadn't taken any time off that year. Peter was always too busy and as the farm was a year-round business, there was work to be done on every single day, including Christmas Day.

'I think a holiday would be a really good idea,' said Mary. 'Once you're fully back on your feet and the docs and have told us you're ready to go back to work, I think we should get away for a while. If we employ a good farm manager, we should be able to leave the farm and it will still be here when we get back.'

Peter hated the idea of holidays. Over the past 10 years or so, he'd barely taken any time off. He sometimes went along to shows and exhibitions but rarely took time away from the farm. Most of his farming friends were the same. They would complain about not having the right staff to enable them to get away and while Peter would never blame poor staff for a reason not to take a break, he simply didn't want to leave the farm for any length of time.

'We'll have to wait until after calving as I can't go away then. It's the busiest time of year on the dairy side. And there's lambing – I can't just leave Andrew while lambing is going on. We then have harvest to consider too. There's always something happening on the farm that I need to be there for.'

'For goodness' sake, Peter. Every year is the same. You make up excuses as to why you can't take a break and I end up going away either with my friends or when the children were smaller, we went without you. When is this ever going to change?'

'It's the farming life though, isn't it?' said Peter.

Mary was getting more flustered and frustrated with the conversation. 'But why? Who said you must work all these long hours and never take holidays? It's no wonder no one wants to join this industry if you never take holidays and practically kill yourself by working all hours God sends. As I've just said, when is it going to change? When you farmers get together it's like a macho competition to see who's worked the hardest. And then when the wives get together, we all complain that we never see you. The only people who are going to change this are us, Peter. We must take the lead and show people there is a way to farm *and* have a life outside of farming. I, for one, never want to ride in an ambulance again like we did yesterday. Our lives flashed in front of me like the blue warning lights. We're going to take a holiday as soon as the new farm manager is in place.'

Peter could sense Mary had made her mind up and admired her courage and wisdom. He also knew she was right.

9

The past 24 hours hadn't been the best preparation for an interview. Martyn hadn't given any thought to it as he'd chased the ambulance carrying his father in a blur of panic and tension. But now he was driving down the A30, heading towards Cornwall and an uncertain future.

For so long, he'd been sure he would run the farm along with his older sister. They would simply agree their individual responsibilities and continue the legacy of their father and grandfather before them.

He had so many new ideas. He liked the idea of regenerative farming or at least moving to an organic system but his father would have none of it. 'Crazy ideas that will just come and go,' was how his father reacted every time Martyn tried to suggest ways to make the farm more sustainable. After attending an exhibition in Birmingham about farm diversification, he'd even suggested using part of the land to build a glamping site. But every idea or suggestion Martyn proposed was met with stony silence and a grim look on his father's face.

As he turned into the farm drive, he noticed a broken sign that said Burton on one half and Farm on the other. The drive was full of potholes and broken gravel and Martyn struggled to keep the car in a straight line. He was already feeling deflated when he spotted a gate on his right, behind which were two rusting and broken tractors. As he continued to be tossed around by the poor condition of the drive, he passed what from the outside looked like a workshop. Inside he could see rusty old tools which were fused together on top of each other. Martyn's heart sank further as he wondered what sort of farming operation was being run by the Burton family.

He knew Burton Farm was a larger, more intensive system where the cows were housed all year, with calving also occurring all year. There were 600 cows and a large team with several Romanian staff who covered the milking shifts, which took place three times a day. As Martyn drove into the farmyard, his spirits lifted as he noticed a brand new, bright green tractor and a few people standing around appearing to admire the large agricultural vehicle. Martyn was transfixed. He recognised the tractor instantly as a John Deere. For so long, he'd been trying to persuade his father to buy or lease a new tractor and every time Peter simply pointed at his trusty red tractor and suggested there were 'many more hours and miles in the old girl yet'.

Martyn parked near two other vehicles. As he clambered out of his tired old Ford Fiesta, he'd already decided he wanted this job. The chance to drive a stunning new tractor like the one in the yard was just too exciting to ignore.

'You must be Martyn,' said a deep voice behind him.

Martyn spun around and looked at the man who was speaking.

'She's a joy to behold, isn't she?'

Martyn was almost speechless.

'Err… yes… it's an amazing tractor. And yes, I'm Martyn. Pleased to meet you.'

'I'm Barry, Barry Burton, farm owner and proud owner of this little beauty.'

Martyn took an instant liking to Barry. He was jovial, friendly and clearly loved new machinery. He was well over 6 ft tall with broad shoulders. He had short, unkempt brown hair with brown eyes and a dark, bushy beard. He reminded Martyn of a pirate from one of his children's books from long ago. Barry put out his large hand and Martyn shook it with as much firmness as he could muster around this giant of a man.

'Come on in, Martyn. Let's have a cuppa before we show you around the farm.'

Barry led Martyn through a door into what was the farm office, although it could have been a dumping ground for old papers as there were invoices, letters, old magazines, some even with their wrappers still on, covering every surface. There were two chairs: one was an old wooden chair with one arm missing and the other was a plastic chair that belonged in a school. The office was certainly not at the same standard as the new tractor outside.

'Excuse the mess, Martyn. I must find time to sort out all this paperwork but it's a bit dull and I keep putting it off. Take a seat and tell me about yourself.'

He pointed to the plastic chair and Martyn sat down, still scanning his surroundings. There was a wallchart calendar from the previous year and a couple of pieces of paper pinned to the wall. The one with 'Calving Protocol' written across the top was curled up around the edges and Martyn figured it had been a while since the papers had been looked at.

'Well, I'm Martyn Wilson and I've been working on my family farm since I was about 14. I run the farm with my sister, mum and dad and I've focused on the cows. My sister looks after the calves and we have a shepherd who takes care of our flock of sheep. I do tractor driving too but Billy does most of the tractor work, so I milk most days, sometimes twice a day, and look after the girls.'

'Sounds good, Martyn. We're looking for you to get stuck into the cow side of things and to run things when I'm not here. Do you think you're ready for the step up?'

Martyn wasn't sure what 'the step up' meant but he didn't want to let on that he wasn't sure. So he agreed that he was.

'Good,' said Barry, who moved over to a table cluttered with about half a dozen cups, a white plastic kettle, a catering-sized box of PG Tips tea bags and a large tin of Nescafé coffee. There was green mould on the surface of a few of the cups covering whatever liquid was still festering within them.

'I must get round to washing up these cups,' said Barry. He took two clean mugs and threw tea bags into each one. He flicked on the kettle and sat down in the wooden chair.

'Let me tell you a bit about our farm then, Martyn. We run a relaxed ship here. If the work gets done, I'm not too fussed about how you do it. Everyone has a slightly different approach to milking and moving cows and I don't mind if people want to do things slightly differently. Adrian, Florin and Marius like to work hard and quickly. They tend to focus on the milking, particularly if they're on the evening shift. Maria, who is Florin's partner, helps with the calves but this is run by Rebecca and Susanna. We then have Shaun, who is our main tractor driver, and Alex, who does a lot of the feeding. I help here and there and make sure everything runs smoothly, including looking after cow health. But I need you to shadow me and then begin to take charge of the farm. Do you reckon you can do this, Martyn?'

Again, Martyn didn't really know what this meant but as Barry seemed quite keen on him, he didn't want to show that he wasn't sure.

'Yes, I'm ready,' said Martyn.

'OK, well drink up. We'll jump in the buggy and I'll show you around although, to be honest, I'm sure when you've seen one farm, you've seen them all.'

Martyn was feeling puzzled. Barry didn't ask any questions about his work or how Wilson Farm was run and any questions he'd asked were easily answered with a yes. It didn't feel like an interview at all. But Martyn liked Barry's relaxed style and he decided to go along with it. He so wanted to be able to drive the big green tractor he knew he'd say yes to almost anything Barry asked just to get the job.

Barry slurped his tea while Martyn put his mug on top of the pile of papers in front of him. He'd barely touched a drop of tea as it was still too hot to drink but Barry was clearly a man who moved quickly as he was already heading for the door.

'Come on, then. Chop chop. Let's show you around Barry's place.'

Martyn followed Barry out of the farm office, across the yard, past the big green tractor and towards a parked Kubota.

'Jump in,' said Barry.

Barry fired up the engine and began to move towards one of the large cowsheds. For the next 20 minutes or so, they got in and out of the Kubota as Barry gave Martyn a whistlestop tour of Burton Farm. By the time they'd returned to the yard, he'd seen all the cowsheds, the milking parlour and a few of the surrounding fields, where maize and grass were grown to supplement the cows' feed.

'So, have you done any AI?' asked Barry as they exited the Kubota.

Again, Martyn knew a little about artificial insemination but they relied on an AI technician to inseminate the cows at Wilson Farm, so he was hoping to get the chance to learn at Burton Farm.

'Yes, I've done a little,' said Martyn, 'but I'd like to do a course to refresh myself, just to bring me up to speed.'

'No problem, Martyn, Adrian does a lot of our AI and I do some too, so we can certainly show you the ropes. So, what hourly rate are you looking for? Or would you prefer a salary?'

Martyn realised Barry was going to offer him the job. He couldn't believe how easy it was, as Barry hardly asked any questions and the job seemed to be his if he wanted it.

'I don't mind really but a salary would be good, I guess.'

Barry was now walking back towards the farm office. Martyn followed.

'The Romanian chaps prefer hourly rates as they can rack up lots of hours and get paid for them, while the rest of the team tend to prefer salaries so they know what income they will receive each month. The choice is yours, Martyn.'

Martyn was already imagining his first drive on the big green tractor and could almost smell its interior. His level of excitement was building and he didn't want to blow it now by asking for too much money.

'How much are you looking to pay?' Martyn asked.

'Does £24,000 plus a mobile home on the farm seem all right to you?'

For the first time, Martyn's enthusiasm began to wane a little. He'd been told that a two-bedroom property came with the job and now he was being offered a mobile home. The salary was slightly less than he was earning on his own farm, too. But he thought about the new tractor and how it would feel to climb in the sparkling cab. He also knew he needed to get away from the home farm to give his dad space. So he decided to accept what Barry was offering but to ask about the house.

'I'm happy with the salary, Barry, but the job advert suggested there was a two-bedroom house with the job?'

'Oh yes, that's the property on the farm I'm planning on doing up,' said Barry. 'It won't be ready for when you start but I'd hope to get it renovated soon after you start. So quite an incentive, eh?'

Martyn could sense flickers of doubt entering his mind. A small voice, barely a whisper at the back of his consciousness, was saying, 'Don't do this'. But the big green tractor... It was so much nicer than all the old, worn-out tractors back on the home farm. He decided to ignore the doubts and recalled his mum talking about trusting his intuition, as she had called it. But as he wasn't sure about intuition or how it worked, he decided to ignore the feeling. He knew he needed to get away from the family farm and this role seemed to be something he could do without too much of a stretch on his part.

'That's fine, Barry – £24,000 and the mobile home would be great.'

'When can you start, Martyn? The sooner the better for me.'

Martyn was keen to get started but knew there was pressure back at home because of his dad's heart attack.

'My dad's not well now,' he told Barry. 'He's in hospital after a mild heart attack so I need to check everything is OK with the family first but hopefully within a couple of weeks?'

'That'll be grand, Martyn. Welcome aboard the madhouse. I hope you're ready for the ride!'

Martyn couldn't believe how easy the whole interview process had been. He'd got himself a new job and a splendid new tractor to drive. He wondered why he'd been nervous at all. There were a few lingering doubts about the salary and the accommodation but the prospect of driving the new John Deere tractor was the sweetener to the deal. He'd longed for the equipment at the home farm to be upgraded but his dad would always point out the increased costs and limited improvements he felt a new tractor would make. But this didn't matter anymore. He couldn't wait to get started.

10

Charlotte was fed up. Once again, the team hadn't been stepping up to the mark. Despite reassurances from Andrew that they'd run a staff meeting to cover rotas, there were still jobs that weren't getting done. Andrew had suggested to the team that they should focus on the essential jobs and had to accept that a few tasks wouldn't get done. But Charlotte didn't agree and told Andrew this wasn't acceptable. She'd told him that just because he was such a long-serving member of staff and the most experienced, this didn't give him the right to suggest that the team could cut corners. Charlotte wanted everything done right and to her standards to make sure that when Peter came back from hospital, the farm was running smoothly. But everywhere she looked, she could see a lack of care: the waste bin at the end of the parlour was overflowing, there were blue gloves on the floor and the washdown was a little slapdash. Based on how they were moving around the collecting yard, she could also see a few of the cows' feet needed trimming.

'Why can't they see these things?' she thought to herself. 'Is it just me who has to go round mopping up after everyone?' She decided to check on the cows she'd already found needed treating. But when she went into the farm office, she couldn't see the diary they used to record all the animals' treatments. It was a strict protocol that the diary shouldn't be moved as it needed to be near the computers for record-keeping purposes. She searched the office but the diary was nowhere to be seen. She turned and stormed out of the office, on the warpath to find whoever had removed it.

She headed down to the cowshed where the sick cows were held. She was particularly concerned about cow 36152. She was an older girl, onto her seventh lactation, but she'd regularly struggled to calve and had endured a couple of doses of mastitis. Charlotte had given clear instructions to the team to keep a close eye on her. Martyn had named the cow 'Sicknote' as she spent a fair amount of time away from the herd being treated but Charlotte thought this was unfair so called her Dottie instead and vowed to take care of her. She had also been present when the cow was born.

Charlotte had asked Amy to check on Dottie at the end of milking and particularly to check her mobility in case she needed her feet trimmed. As Charlotte walked down towards the shed, she could hear unusual noises that quickened her pace. By the time she'd reached the shed doors, she could see Dottie was lying on her side in the large pen with the sick cows.

'Oh no,' thought Charlotte. 'Where is Amy?'

She opened the gate into the pen and walked over to Dottie. Steam was rising from the cow's body. 'I bet she has an infection!' Charlotte thought. She grabbed her mobile phone and looked for Amy's number. She dialled it and as soon as Amy answered, she bellowed down her phone.

'Where the heck are you, Amy? Dottie's sick. Did you check on her this morning like I asked?'

There was a pause on the end of the line before Amy responded. 'No, George was supposed to check the sick cows. I've been helping with milking this morning.'

'But it's not good enough, Amy,' shouted Charlotte. 'For heaven's sake, can't I leave you to do anything?'

'I'm sorry, Charlotte. Shall I come over now to help you?'

'It's too late for that. I'm going to have to call the vet. Just leave it. I'll sort it.'

Charlotte ended her call with Amy and dialled Tom Storey's number. Thankfully, he answered at once.

'Hi Tom, it's Charlotte Wilson here. We have a cow down that looks very unwell. She's already been treated for mastitis but isn't responding. Could you come over to us ASAP please?'

After discussing the treatments administered to Dottie so far, Tom agreed to the visit. Charlotte hated seeing sick animals. While the protocol on the farm was to use antibiotics cautiously, they recognised their value when needed. Dottie was a special cow and Charlotte had helped at her birth and nurtured her throughout the first phases of her early life. When she was in calf for the first time, Charlotte felt a huge sense of pride, and after that she took a keen interest in Dottie's pregnancies. And now, due to staff incompetence, Dottie was in danger of losing her long battle to stay alive. Charlotte went over to Dottie, who appeared to be restless and uncomfortable.

'How are you, Dottie?' asked Charlotte. 'I'm sorry this hopeless lot have let you down. Help is on the way. You just hang in there, OK girl?'

Charlotte stroked Dottie's nose. She lifted her head to look up at Charlotte and, as if reassured, she put her head back down onto the bed of straw where she lay. The natural curiosity of the other cows in the shed ensured they weren't short of company while they waited for Tom to arrive. For more than 30 minutes Charlotte stayed with Dottie and when she heard Tom's car coming down the farm drive, she felt relieved.

Tom put on his overalls and boots. Stepping into the dip outside the shed, he moved through the small number of cows to find Charlotte with Dottie.

'Hi, Charlotte. What's up?'

'I'm certain it's another dose of mastitis. She's hot, her front quarter is red and she's just starting to show signs of diarrhoea and dehydration. I'm really worried about her, Tom. I can't lose her. I've put too much into this girl for us to lose her now.'

Tom examined Dottie and explained what he intended to do. Charlotte watched Tom as he administered antibiotics and suggested ongoing treatments for the team to give to Dottie.

Charlotte listened carefully but was finding it hard to concentrate. The emotions of the last few days were creeping up on her and she was feeling out of control. This was uncomfortable for Charlotte as she liked to be in control of everything, including her feelings. But now, with her father in hospital, the team not performing and her favourite cow struggling to hold on to life, she couldn't hold back the tears. She squeezed her eyes shut, hoping this would prevent tears from leaking out but to no avail.

'Are you OK, Charlotte?' asked Tom.

Charlotte sniffed. 'Yes, Tom. Just been a rough couple of days with Dad going into hospital. And I'm so frustrated with the staff. No one seems capable of doing anything I ask of them. I just feel as if I'm the only one who sees what's going on and what needs to be done. To be honest, it's exhausting. And now Dottie is poorly too. It just pushed me over the edge but I'll be fine. You get Dottie well and I'll be happy.'

Tom held Charlotte's arm. 'Sometimes just admitting it's all too much is the first step to getting back in control. I struggled with my mental health a few years back and it was only when I finally admitted I was struggling that things started to improve.'

Charlotte was shocked. Tom was one of the most 'together' people she'd ever met. It never occurred to her that he may have struggled, let alone have mental health issues.

Tom could see from Charlotte's expression that she was startled by his admission but he could also see she was teetering on the edge of a full meltdown. He'd seen it before on other farms and with colleagues. They thought they were the only ones capable of doing anything to the right standard and would then find themselves utterly exhausted and burnt out. It was an all too familiar pattern within the farming industry.

'Have you spoken to anyone about your dad's heart attack, Charlotte? Or your concerns about the staff?'

Charlotte shook her head. 'I don't want to bother Mum as she's got enough on her plate and Martyn is off today at an interview. The rest of the staff hate me so there's not much point in talking to them and since Young Farmers closed its meetings due to the pandemic, I've not really found anywhere to go to offload.'

Tom knew how Charlotte felt. 'When I was struggling, I called the Farming Community Network and there are lots of other great charities out there who can help too. Even the Samaritans will listen if you're having a hard time. You don't need to suffer in silence or bottle it up, Charlotte. Maybe Dottie getting sick is the final push you need to recognise that you're not Wonder Woman Wilson. You're a normal, young human being trying to do your best in very stressful circumstances.'

As Tom began to pack away his things and remove his overalls, they chatted more about the strains of being in a farming family. Tom explained that his parents split up when he was a teenager studying at agricultural college and how he took to drink to ease his pain. When he finally qualified as a vet, he hoped his days of feeling low, trapped or unworthy would end but they didn't. And when a farmer blamed him for losing one of his cattle, Tom descended back into a pit of despair he felt he may never emerge from. He admitted to dark thoughts of taking his own life and it was only another vet at the practice admitting his own mental health struggles that encouraged Tom to finally seek the help he needed.

Charlotte listened intently to Tom's story and while she was nowhere near the depths of despair in which Tom had found himself, listening to his journey helped her to see she needed to speak up about how she was feeling. As Tom drove away from the farm, Charlotte turned back to check on Dottie. She was still lying down and appeared to be slightly calmer than before. When Charlotte approached her, Dottie turned her head and made a brief noise as if to reassure Charlotte she was going to be OK. Charlotte kissed her hand and then placed it onto Dottie's head.

'If you can get through this, Dottie, I can get through this tough period. You're a strong girl, so let's do this together, yes?'

Charlotte may have imagined it, but she was certain Dottie nodded her head. 'You're a special girl, Dottie,' she thought. 'So are you,' was the response Charlotte heard in her head.

11

'Hello, Mary, it's James Glover. I just wanted to check you're OK for my visit on Friday to run the Team Dynamics workshop. After Peter's heart attack and with the extra pressure you must be under, I wondered if you wanted to postpone the session?'

'No, James. Let's press on,' said Mary. 'I want to be clear on the type of person we need to run the team and from what you said last time, this workshop will enable you to help us with this. Let's do it. Peter might be home by then so would be able to join us too. What time will you be here and what do you need from us?'

'I'll be with you by 9.30 am for a 10.00 am start, which should give everyone time to complete morning milking and have breakfast,' said James. 'We just need a room big enough for us all to sit down, ideally with a table so we can write things down. I'll bring all the materials so all you need to provide is the tea and coffee – and maybe biscuits!'

'That's fine, James. I'll get it all organised, so we'll see you on Friday.'

'Sure, see you then.'

Mary was looking forward to the workshop James planned to deliver and decided to catch up with paperwork. She had suppliers to pay and feed to purchase, so was keen to get on. James's call was a timely reminder that the recruitment of the new farm manager was imminent. She found herself thinking about the Chairman and while she didn't really know why she felt compelled to call him again, she trusted her intuition and dialled his number.

'Hello, Mary, how are you and how is Peter?'

Mary was taken aback. How did David know about Peter? She assumed he'd heard through their network or other farmers.

'Peter is doing well, thanks, and should be home tomorrow or the day after.'

'Good news, Mary. And how are Charlotte and Martyn after all the upset?'

Mary didn't have time to consider how they'd been coping. She'd focused her time and energy on keeping the farm going. This, along with knowing Andrew and Billy were holding the team together, meant her children had been left to their own devices.

'I think they're OK, although I've been so busy, I've not really focused on them,' she continued. Then there was a silence. David appeared to enjoy or utilise silence when he needed time to think.

'Big situations can make us or break us, Mary. And sometimes a breakdown comes before a breakthrough. We often need to admit we're not coping for other people to be able to help and for us to stretch and grow. It's often a confusing paradox. People can believe the tough times in life are to be avoided at all costs. But if we look back on our lives, it's always during the tough times that we find new strength, heal relationships or start new ventures, isn't it? People and circumstances that appear synchronistic or meant to be can enter our lives and then random events often start to make sense – particularly if we look back in hindsight.'

Mary was thinking of the tough situations that she and Peter had faced during their marriage and time on the farm and she realised they had always emerged stronger and wiser.

'It's all about endings and beginnings,' said David. 'Often, what we thought was the end of something was just the beginning of something else. Watch out for the sequences, Mary. You'll be amazed how life begins to make sense even when, at first, it doesn't seem to.'

'I hope you're right, David. In fact, I just called James Glover and he's coming to the farm on Friday to run a workshop for us. I suspect that if Peter hadn't suffered his heart scare, he may not have been interested in James coming to the farm. He'd have said he was too busy to take part. So, you're right. There does seem to be a sequence to events.'

'Yes, Mary, and if you stay open minded and willing to accept this time on the farm as a breakthrough rather than a breakdown, then the future may be very different to the past.'

'I hope so, David.'

'One more thing, Mary. I sense you may need to spend time with your children to check in with them too. When a business is going through a difficult time, it's often the family who pick up the slack while also taking on the pressure.'

'I will, David. I'll see them both today and make sure they're OK.

Thanks again for taking my call. Your perspective is always reassuring.'

Mary didn't really know why she'd called the Chairman and realised he'd given her an answer to a question she'd not asked. As well as being aware of the synchronistic events that might occur, David's invitation to focus on her children was a painful but welcome reminder to put the family before the farm. She'd spoken to all the staff, made sure suppliers were being paid and the cows fed but she'd not spared a moment to ensure her family members were OK. It was a frightening realisation.

After the call, Mary reached for her notepad. She turned to the page where she'd been making notes and added three more.

Find the silence.
Recruit in haste – regret at your leisure.
Read between the lines.
Trust your intuition.
A breakdown is often a breakthrough.
Watch for the sequences.
Put family first.

Mary reviewed her notes and while much of what the Chairman said didn't appear to relate directly to the farm, she realised his guidance was deeper, broader and appeared to be about life as well as business. She turned on the office computer to check her emails and pay some suppliers. Mary wondered who their farm manager might be. Somewhere out in the world was a person who could change their fortunes. As she waited for the computer to load all its software, she sat back in the chair in silence. Other than the sound of electricity and the distant hum of cows and farm machinery, the office was completely quiet. She decided to use the silence to ask herself a few questions, just as David had suggested.

'Will we find a farm manager?'

'Yes,' came the reply. Mary knew it was her voice but somehow it didn't feel silly to be talking in her mind. Growing in confidence, she decided to ask her silence another question.

'Who will this person be?'

'A strong, independent woman who is keen to learn, experienced in dairy farming and brilliant with people.'

Mary wasn't sure if this was just her imagination but as she heard the words in her head, it felt real. 'Maybe I'm just hoping for a strong woman,' she thought. She also knew farming was still very male dominated and the chances of finding a female farm manager were slim. It was far more likely they would be a male and Mary fully accepted this.

Her chat with the silence was interrupted by her mobile phone, which was vibrating and flashing. She looked down to see it was James Glover again.

'Hello again, James. What can I do for you?'

'Hi, Mary. Yes, sorry to call again but I wanted to let you know I'm sending a CV for you to look at. It came in a few minutes ago and although we need to profile the team at the workshop to see how this person would fit, her background looks interesting.'

'Her?' Mary thought.

'I can see from her experience that she's been a herd manager on a large organic dairy unit and has experience of working with sheep. I'll check out her references today but, in the meantime, could you send over the job description for the role and your person specification? Then I can see how she matches up against those documents.'

Mary was still coming to terms with the fact that, moments before James called, she was talking to the silence about a woman being the next farm manager. 'OK,' she thought, 'so it was 50/50 as to the gender of the potential candidate.' But it did feel a bit like one of those synchronicities David was talking about. Mary also realised she didn't have a job description for the role – or indeed any other role on the farm.

'That all sounds good, James. We don't have job descriptions, I'm afraid, and I'm not sure what you mean by a person specification.'

'No problem, Mary. We have a bank of generic job descriptions. Based on what we talked about for the job advert, I'll prepare one for you to look through and you can make any tweaks as you see fit. It's important to have a job description for the role you're recruiting for, as it helps to identify the skills and knowledge any candidate may need to be able to demonstrate. Plus, how will you ever know if the person is fulfilling the role effectively if you don't know what you're expecting them to do? It's also important we have an organisation chart to show the potential candidate how the role fits into the farm and the reporting lines, so I'll prepare one of these for you too.'

'That would be fabulous. Thanks, James. And the person specification?'

'Ah yes. The job description states what the person will do on the farm while the person specification is about who they are. By this, we mean the character, personality style and possible qualifications they may need. We'll know more about this after our session on Friday. So again, I'll send something over for you to look at.'

Mary was a little overwhelmed by how helpful James was being – but Anna did say REAL Success were amazing.

'That's fabulous, James. I look forward to seeing all those documents and in particular the CV of the lady you mentioned. What's her name?

'Claire Davis,' said James. 'I'll send over her details now. Speak soon, Mary.'

As Mary put down her phone, she sensed that something was beginning. She couldn't work out if she felt nervous or excited. Or was it both? Her computer flashed a few times as her emails opened. She clicked on the new email from James with Claire's CV attached. She knew the job description and person specification would come later. She clicked on the attachment and opened Claire Davis's CV and covering letter. As James had suggested, she was currently working for a well-known organic farm with three dairy units across Devon, Somerset and Cornwall. She was a herd manager on the largest unit and had been there for more than seven years. She'd started as a general farm worker, progressed to a herdsperson role and went on to become herd manager. She'd clearly been involved in growing grass and crops and managed a team of six. Prior to working on the dairy farm, she'd been a shepherd on a large sheep farm. Her covering letter said she was now looking for an opportunity to grow her experience and manage either a multi-site unit or a mixed farming enterprise.

As Mary scanned her CV, a shiver went down her spine. Claire was exactly what she'd imagined when she was talking to the silence. Was this the intuition David talked about? Or was it just a pleasant coincidence?

'Morning, Mum,' said Martyn as he came into the office.

'Hi, love,' said Mary, still staring at Claire's CV. 'How are you this morning and how did the interview go yesterday?'

'Oh Mum, it was awesome. Barry showed me round the farm and while it doesn't seem to be as well organised as we are here, he's just bought this huge new John Deere tractor. You should see the technology inside the cab, Mum. It's like something out of a spaceship. The GPS system is brilliant and there were so many features, it would take me a month to work them all out!'

Mary wasn't surprised Martyn was impressed. He'd been saying to Peter for months that they should invest in a new tractor. However, she wondered about the rest of the farm and the role.

'How about the cows? Were they in good condition? And what's the role he wants you to fill?'

'The cows looked OK. The sheds were a bit mucky for my liking and a few cows appeared to be lame but I'm sure they'll get on top of that. And the role is cool. It's a second in command position so I'd be the senior herdsperson and managing the farm whenever Barry was off the farm.'

'And the accommodation?'

'There's going to be a two-bedroom house with the job. It's not ready yet so initially I'll be in a mobile home on the farm but Barry promised to get the house sorted ASAP.'

Mary's mood changed. She felt a sense of foreboding. It was a dark, worrying feeling.

'Are you sure this farm and this farmer are right for you, Martyn? It all sounds a bit slapdash to me. And I thought you said the job advert offered a property with the role – not a mobile home. You've got a nice house here with us so why would you want to live in a mobile home?'

'I know, Mum, but Barry was keen for me to join and although the money is a little less than what I take from our farm, I think the opportunity is a good one. And he's really open to new ideas too. I spoke about a few things and his philosophy is to allow the team to do what they want as long as the work gets done. He's the opposite of Dad in that respect and I like the idea of having the freedom to make decisions around the farm. And Mum, you should see this new tractor – it's totally awesome.'

Mary still felt uncomfortable but Martyn seemed keen. With everything going on, she didn't feel inclined to challenge him further. Instead, she decided to follow David's advice and ask Martyn how he was dealing with all the changes happening around them.

'I'm fine, Mum,' said Martyn, trying to reassure his mother. 'I like change and while I wouldn't have wanted Dad to end up in hospital, things are moving forward now. It's weird. It's like all these things were meant to happen. Yes, I'm a bit sad to be leaving the farm at this point but I still think it's for the best for all of us. I've told Barry I could start in a couple of weeks – is that OK? He's keen to get me on board as quickly as he can.'

'Crikey, that's a bit quick, Martyn. I'm hoping to have the new farm manager in soon and an excellent CV has come in this morning but ideally, could you not stay for another month?'

'Not really, Mum. Barry has plans to expand his farm and he wants me there from the start. Charlotte will be here and Dad will be back at work soon, won't he? I really want to start as soon as possible. Plus, I've never been given a proper contract so when I checked online, the minimum notice I must give to you and Dad is one week. So two weeks is quite generous, I think.'

Mary was slightly stung by Martyn's lack of concern for the farm, his family or his colleagues, who would all have to step up and take on part of his work.

'Two weeks until what?' asked Charlotte, who had just come into the office. Mary thought she looked as if she'd been crying.

'Are you OK, love?' asked Mary.

'Yes, I'm fine. Why? And what's happening in two weeks?'

'Martyn has a new job and starts in two weeks.'

'Two weeks? Blimey, Martyn! Can't you see the mess we're in right now? We nearly lost Dottie again yesterday because the staff can't spot and treat a simple case of mastitis, Dad's in hospital and you think this is a great time to go off and do your own thing? Thanks for nothing, brother.'

Forever the peacemaker, Mary didn't want the children to be squabbling when everyone needed to look forward.

'We'd all agreed that Martyn was going to look for a role on another farm and it sounds as if the farm he's going to will be ideal for him.'

'It's a cracking farm, Charlotte,' said Martyn. 'They've just bought this amazing new tractor and the farm owner has really exciting plans for the future.'

'Oh, a new tractor. Well, I can see how that seals it for you,' said Charlotte sarcastically. 'Can't you see we need you here? *I* need you here? Just put your differences with Dad to one side and stay and help us build something.'

'But you're advertising for a farm manager now, which restricts my chances of growth *and* yours Charlotte. I want to spread my wings and try something new and opportunities like this don't come along very often. I'm afraid I'm leaving in two weeks.'

'What's happened to Dottie?' asked Mary, keen to change the subject.

'She's OK now,' said Charlotte. 'I found her lying down and overheating so I called Tom and we've sorted her out. Just like everything else around here, it was left to me to sort things out.'

And with that, Charlotte turned around and left the office.

'I think my sister is a control freak,' said Martyn as the door closed. 'She thinks no one can do anything on this farm as well as she does. And Dottie has endured more comebacks than a pop group who have seen better days. She always survives so I don't know why she's so upset about just one cow.'

'Dottie isn't just a cow to Charlotte, Martyn,' said Mary. 'She was the first calf she brought into the world and she's always been a little sickly. I think, in a way, she feels responsible for her, which is why she gets upset when others in the team aren't looking after her.'

Martyn shook his head. 'I think she cares about the cows more than she does the rest of the team. She's awful with the staff, Mum – you'll have to watch her when I leave.'

Martyn left the office and headed over to the ageing tractor his father refused to replace. As he clambered into the cab with its broken switches and rusty interior, he wondered how it would feel to climb into the tractor at Burton Farm. His mind was made up and he couldn't wait to get started.

Mary was troubled by her children's bickering but wondered if they too would benefit from time apart. She looked back at her computer screen and read through Claire Davis's CV once more. In her personal details, she referred to a partner but no children. Mary wondered if Claire's partner also worked in farming. With farming families, couples would often both work within the industry and this would mean housing would need to be suitable for a family. Claire's CV also referred to two Labrador dogs called Tilly and Max. While Andrew owned a sheepdog called Bertie, the Wilson family didn't have dogs. This was quite unusual in farming as most farmers owned at least one dog and those who also owned land used for pheasant shooting would always have several dogs for the shooting season.

The office door opened and Marek appeared.

'Sorry to come in, Mary, but can I have a quick talk, please?'

'Sure Marek, come in, sit down.'

'Thank you, Mary. I just wanted to tell you my mother has died in Poland. I must go back for a funeral next week and I need to stay in Poland for a while to make things good. You know?'

Mary's heart sank – not only for Marek's loss but it would also mean another member of staff would be leaving the farm, making the team even tighter on resources.

'I'm so sorry, Marek. Was your mother ill or has this been a shock? You didn't mention anything about your mum being unwell?'

Marek looked tired and his eyes were swollen as if he'd been crying. 'No, it's a complete shock. A heart attack, like Peter. But my mother was not so lucky. She died in front of my father who is also not a well man. He has cancer of the stomach and my mother was caring for him. He is now in hospital too as my family cannot look after him.'

'Of course, Marek. You must fly home straight away and be with your family. How long do you think you'll be there before you can come back to us?'

Marek looked at the floor. 'I'm not sure if I be able to come back, Mary. My brother and sister in Poland live far away from the hospital where my dad is being treated and I must look after the house and funeral and all these things. So, I can go for a month but will call when I know what my family need from me. Is this OK, please?'

Mary's mind was racing. She would need to find another member of staff as Marek was one of the key milking staff. He was reliable, hardworking and caused very little fuss around the farm. Finding people like Marek wasn't easy and ever since the UK left the European Union, it had become even harder to find good quality overseas workers to perform the basic tasks around the farm.

'Of course, Marek. You must put family first. Please don't worry. Just tell me when you've booked your flights, OK?'

Marek stood up and offered Mary his hand.

'Thank you, Mary. Wilson family have been good for me and my family, so I try hard to come back, OK?'

It occurred to Mary that during the past few hours, she'd been dealing mainly with family issues. She glanced down at her notepad and saw the final note in her list after her last conversation with the Chairman.

Put family first.

Mary smiled. He'd been right, of course.

12

Peter had been ready to come home since 4 am. So, when Mary appeared he breathed a sigh of relief.

'Morning, love,' said Peter as Mary approached him across the ward.

'You look keen,' said Mary.

'I'm just keen to get home and get back to work. What's been happening on the farm? Is everything OK?'

Mary wondered where to start as the past few days had been a whirlwind.

'Everything is fine, love. You'll need to take it easy for a few days anyway so no leaping back into a tractor for you, I'm afraid.'

'But farming is all I do, Mary! I'll just get bored sitting around doing nothing. I'll be fine.'

Mary wasn't convinced and started to unpack the bag of fresh clothes she'd brought for Peter to change into.

'Have they discharged you, then?'

'No, not yet. I'm just waiting for the forms and letter for my GP. The sister on the ward reckoned I should get them soon. Then we can go.'

Mary laid out the clothes on the bed and started to draw the fabric curtains so Peter could change out of his hospital gown. As Peter began to get dressed, Mary decided to update him about what had been going on back at the farm.

'I was catching up with Martyn and Charlotte this morning.'

'Oh, yes,' said Peter. 'Are they OK?'

'Martyn is, yes; Charlotte isn't,' said Mary.

'What do you mean?'

'Martyn is happy as Barry Burton offered him a role on his farm. I'm a little dubious to be honest as we know Barry has been through a lot of staff who all say he runs an untidy farm where cow welfare isn't high on the list. They might produce more milk than we do but the working conditions aren't brilliant. But he's bought a new John Deere tractor, which of course sealed it for Martyn. So he's given us notice and he's off to work for Barry.'

As he buttoned up his shirt, Peter looked up at Mary. 'Burton Farm isn't a good farm but I guess if Barry has offered Martyn the crown jewels, there isn't much we can do. And a new tractor was always going to turn Martyn's head as he's been on at me for months about replacing ours.'

'That's the crazy thing. The advert suggested there would be a house with the role but it's a mobile home and the salary is slightly less than we're paying him. Honestly, I think it was simply the new tractor that clinched it for Martyn, so he leaves in two weeks.'

'Blimey, that's quick. But I guess we *have* talked about it for a while so we can't hold him back now. I'll chat to him when we get home.'

'And Charlotte seems a bit down, Peter. I'm worried about her. She gives the impression that she's strong and decisive but she was really upset this morning as Dottie is poorly again. She thinks no one else on the farm cares about the animals and the farm like she does.'

'Is Dottie OK?' asked Peter. 'Charlotte's always worried about her.'

'We think she'll be fine – it's just a nasty case of mastitis again but Charlotte is right. One of the team should have spotted that she wasn't well. Charlotte is also furious with Martyn for abandoning us at such a tricky time.'

Peter was fully dressed and sat back down in the chair next to his bed. 'He's not abandoning us, Mary. He's just spreading his wings and I think we need to let him do it. One day he'll come back to the farm but if we hold him back, he'll just resent us and the farm. And that's not good for the long term, is it?'

Mary knew he was right. Martyn needed to try working on another farm to compare with the family farm. 'But Charlotte isn't coping, Peter. I think we need to keep an eye on her. We can't have her breaking down and not being able to work as we're going to be short staffed anyway.'

A thought flashed into Mary's mind. It was something the Chairman had said about a breakdown being a breakthrough. She wondered if Charlotte needed to recognise that she couldn't do everything around the farm and would then understand the importance of building relationships with the staff. 'Perhaps her emotional breakdown could be a breakthrough for her?' Mary thought.

'Why are we going to be short staffed?' asked Peter. 'Is someone else leaving too?'

Mary told Peter about Marek and they both agreed that while it would create a big hole in the milking team, he should, of course, put family first and head back to Poland.

'But in better news, James Glover has sent me the CV of a potential farm manager.'

Peter screwed up his face. 'Do we really need someone to come in and tell us what to do? I'm perfectly capable of running the farm.'

Mary flashed a look at Peter. 'We've been through this already. We need help. I need help. It's not all about you. I'd like to be able to talk through the finances of the farm and share some of my concerns about how we spend money on things that don't seem to add any value. Plus, I don't want to travel in an ambulance with you again.'

Peter knew she was right but still felt uncomfortable about having someone else run the farm. Farming was all he knew and if he wasn't making the decisions, what would he do all day?

'OK, love. But I don't want to be a spare part on my own farm. I've worked for years to build what we have and I don't want a young upstart coming in and changing everything. That's all.'

Mary didn't reply and for a few moments there was an uncomfortable silence as Peter gathered his belongings.

'You're good to go, my lovely,' said the ward sister, poking her head through the blue fabric curtains. Seeing that Peter was fully dressed, she stepped through the gap and stood next to his bed holding papers in her hand. Sister Joan was a small woman who Peter estimated was in her sixties and nearing retirement. With silver hair pulled back into a bun, she had the weather-worn face of someone who'd spent a lot of time outdoors. She was from North Wales but had worked in poor areas of the world as a volunteer for the Red Cross, often coming across farmers who had very little compared to those in the UK.

'You mind what I told you this morning then, Peter. There's no point in pushing yourself to do more and more. It can be better to aim for less and be happy with what you have.'

Mary's ears pricked up as she heard what Joan had said.

'I've told him so many times, sister. His father used to say it all the time too. He'd tell him to aim for less all the time. So many farmers I meet just think the answer is to have bigger farms, more cows, larger sheds, bigger tractors. I've told him we have enough already and although he agrees with me most of the time, I worry about what his farming pals and his farming consultant tell him.'

Joan cocked her head towards Mary. 'You need to listen to your lovely wife, Peter. We don't want to see you back here anytime soon so take a good, long look at your lifestyle and make sure you've got enough help around you. You hear me?'

Peter felt he was being bullied into submission by the two women standing over him, but deep down he knew they had a point.

'Ok, sister, I hear you. I'll be a good boy and do what I'm told.'

Sister Joan laughed. 'That's the ticket. Now here are your discharge forms and this is the letter you need to give to your doctor. They'll want to keep an eye on you, so you'll have to go to the surgery for an ECG and a general check-up in about a month. Now let this lovely lady take you home and no leaping straight back to work, OK? Take a few days off and ease yourself back in. Your ticker has given you a warning, Peter, and you'd do well to listen to it. Nice to have met you, Mrs Wilson, and in the nicest sense of the words, I hope we don't see you again soon.'

As Joan walked away, Mary said, 'Come on love, let's get you home. It's time to start a new chapter in our lives. I'll tell you in the car about the person James Glover has put forward for the farm manager position. She sounds interesting.'

'She?' said Peter. 'I don't think I want another female telling me what to do. I have enough with you and Charlotte. Anyway, who's ever heard of a female farm manager? I can't imagine she has much experience.'

Mary bristled at Peter's attitude about having a woman lead the farm. As they headed towards the ward exit and walked down the long hospital corridors, she questioned him about his reaction.

'I like to think I'm broad minded,' said Peter, 'but women have babies. She'll work for us for a little while, get pregnant and then I'm left funding maternity pay and having to either cover the position myself or employ another person. Surely we'd be better off with a bloke running the farm?'

'That's the problem with this industry, Peter. To outsiders, it must seem that you need to be white, British and male to lead a farm, which doesn't reflect the society we live in anymore, does it? Farming is way behind other industries and I think it's time we started to reflect the world in which we live. I'm afraid it's only women who can produce offspring so unless you want population control, you must accept that women may have babies. Anyway, you're making so many assumptions already and we haven't even met the woman yet.'

By the time they returned to the farm, they'd debated the gender balance across farming along with the lack of ethnic diversity and why this may the case. They even briefly discussed the lack of lesbian, gay and other types of gender diversity within the industry. While Peter didn't agree with the need to positively discriminate to rebalance the bias towards white, British males, he did accept there could be prejudice within the industry that needed to be broken down.

'Yes, but a breakdown can often be the breakthrough, Peter,' said Mary. 'Maybe we can show Charlotte she doesn't need to try so hard to be like her male colleagues and having a woman running a farm is perfectly acceptable.'

As Mary turned the truck onto the farm drive, Peter gave out a gentle sigh.

'It's good to be back,' he said. 'I've missed this place.'

'You've only been away for 48 hours, for goodness' sake! Anyone would think you've been away for a month.'

They laughed together and it felt good. Mary wondered again whether this was the dawn of a fresh start for them all. Peter simply wondered who'd been milking, how Dottie was recovering and how lambing was going…

PART TWO

The Team Develops

13

'I don't know why we're doing this workshop thing today,' said Billy, who looked fed up. It was only 9 am.

'I'm not sure either, Billy,' said Andrew, who was equally nonplussed about the meeting they were about to attend.

As they walked across the yard towards the house, they continued to mutter and complain about being asked to attend a team-building session with this chap called James Glover.

George was much more positive about the meeting. He enjoyed new experiences and could see the benefit of everyone understanding more about each other. Marek had agreed to come along, despite his family news and his plan to fly back to Poland the next day. He was already sitting at the kitchen table looking slightly impatient as the rest of the team began to filter into the house, removing their boots and leaving them at the back door. Amy felt nervous that the session was going to show her up in front of her colleagues.

The smell of sizzling bacon filled the air and Mary was busy making tea and coffee for everyone. Peter sat at one end of the kitchen table while the stranger they all assumed was James Glover was busy connecting his laptop to the TV on the kitchen wall. Charlotte sat next to her father and as everyone else sat down, it was a tight squeeze to get everyone around the table. The conversation quickly turned to Peter, his heart scare and why he was back on the farm so soon.

'Aren't you going to take some time off, boss?' asked Marek.

Peter reassured everyone that he was intending to take it easy for a few days, as the doctors instructed, but he was fit and well enough to continue light duties around the farm.

'Where's Martyn?' asked Charlotte.

The team all looked at each other, shrugged their shoulders and looked puzzled.

'I think I saw him at the bottom field moving fences when we brought the cows in this morning,' said Amy. 'But I could be wrong.'

'He's always late,' said Charlotte. 'Does my head in. Why can't he be on time for once?'

'He'll be here in a minute or so I'm sure,' said George.

Mary placed a large plate of bacon sandwiches on the kitchen table and everyone tucked into them gleefully. The chatter died down a little as they began to eat, so Mary decided to introduce James.

'While you're all eating your sandwiches, let me introduce James Glover from REAL Success. James is going to help us to find the farm manager we talked about last week but he also helps teams perform better by understanding each other more deeply. I thought it would be useful for us all to get to know each other better, particularly with the changes in the team, with Martyn leaving soon and Marek heading back to Poland tomorrow. James, would you like to take it from here?'

James stood up at the far end of the table where Peter and Charlotte were sitting. He looked at Mary. 'Should we wait for Martyn to arrive? I think it's important for him to be here.'

Then the door from the kitchen to the small utility room burst open and Martyn nearly tripped over Mary, who was standing by the sink.

'Morning, everyone! Sorry I'm a bit late. Got distracted by a fence that needed fixing.'

Charlotte looked furious while Andrew smiled. Various knowing glances were exchanged between the team, who weren't at all surprised Martyn was the last to arrive.

'Sandwiches look good, Mum,' said Martyn as he slid into an empty chair.

James started to speak once more. 'Good morning, everyone. My name is James Glover and my company is called REAL Success. Thank you all for coming along and I'm sure you're wondering why on earth you've been dragged inside to participate in this workshop.'

Andrew and Billy shifted uncomfortably in their seats while James went on to explain a little more about his background and outlined the services that REAL Success offered. He then asked the team to offer suggestions for the single biggest reason people leave a farm to work elsewhere.

'More hours and more money?' offered Marek.

'Career advancement?' said Charlotte.

'Retirement?' offered Andrew, and everyone laughed.

'Better working conditions,' said Martyn through a full mouth of bacon sandwich.

'Family disputes?' suggested Mary, who was still standing next to the sink.

James nodded. 'Yes, all those things are factors but there's still one that no one has mentioned, which is the biggest reason. Any further thoughts?'

'Me,' mumbled Peter.

'Ah yes, Peter,' said James. 'I'm afraid you're right. The single biggest reason people leave a farm is because they can't get on with or don't like their boss or a colleague. So you are all the biggest reason staff might leave.'

Charlotte glanced at Amy, wondering if she'd been a little harsh on her. Billy and Andrew looked at each other, each thinking about their conversations about Charlotte. Marek was thinking about the previous farm he'd worked at and how the boss was lazy and the farm a mess. Martyn stopped eating his sandwich and looked across at his father. Mary was observing everyone and knew James was right.

'So if you're the reason people are leaving the farm, what can we do to prevent this, or at least reduce the possibility of this happening?'

'Get Dad to retire?' offered Martyn and once again, everyone laughed, which eased the slightly awkward atmosphere.

'Maybe if we just learned to understand more about each other?' suggested Mary.

'Well, Mary has it right,' James responded. 'If personality clashes are the single biggest reason people leave a farm, then if we understand a little more about our personalities, we might find we clash less, are not so offended by the behaviours of others and might just be able to resolve our differences. You see, it's our differences and not our similarities that are our strengths as a team. And by the end of this session, you'll be able to spot the differences between you all and then work on adjusting how you work together to be even more productive, even more efficient, much happier and welcoming of new ideas.'

Peter had already decided he liked James Glover. He was clear and concise and Peter sensed that what James was going to cover would enable everyone to have a better understanding of each other, leading to greater efficiency around the farm.

Charlotte was frustrated and wanted James to get to the point. They'd been talking for about 10 minutes and she'd learned nothing new. She wondered what the result of this workshop would be and if Amy was going to buck up her ideas. Amy liked James. He seemed friendly, warm and to genuinely care about helping the team. She hoped Charlotte was listening to what he said and was going to soften her approach a little. George was bored. He'd noticed the postman was about to deliver a parcel. 'Postman's here,' he said, interrupting James.

James continued regardless.

'So this morning's session is about helping us notice our personality differences and how this affects how we communicate with and behave towards each other. We all "ooze clues" and by the end of this session, you'll be able to spot these differences and know what they tell us about the person you're working with.'

Andrew scratched his chin. He was reflecting on what James had said and could remember a number of people who'd worked on the farm and who'd clearly been very different to him. Andrew was specific about how he looked after the sheep, with a set routine he followed on most days. He recalled a young lad called Henry who joined the farm but only lasted three months. He'd worked in a chaotic manner, was always suggesting new ideas and would forget what he'd been asked to do. He reminded Andrew of Martyn and Billy.

James handed out small, pre-printed notebooks and gave everyone a pen with the REAL Success logo printed on it. He outlined why he'd supplied notebooks for everyone and asked them to turn to page two, where a list of numbers under the heading 'Active Listening' was printed.

'To illustrate an important point about communication, I'm going to ask you ten simple questions,' said James. 'Don't ask me to repeat the question, don't copy the person next to you and don't giggle.'

Amy let out a giggle and everyone else chuckled too. James started to ask the group questions and, very quickly, there were puzzled faces and confused expressions all around. The occasional sigh or giggle could be heard as the team grappled with the strange questions James was asking.

Once James had finished asking the ten questions, there was an abrupt end to the peace and quiet and everyone started talking to each other about how difficult it was to answer them.

'I'll let you mark your own papers,' said James, 'so let's go through them one by one.'

James then asked each question again and gave the correct answers. In every case, the answer the team expected to be the correct one wasn't, and it became clear that in most cases the team simply 'misheard' the question or didn't realise there was a trick in the question. The team laughed and groaned as they discovered that even though the questions were simple, they'd got most of them wrong. James asked them to add up their scores and Andrew was declared the winner with three correct answers out of the 10 questions asked.

'I knew they were trick questions,' said Andrew, 'so I just listened carefully to what you were saying. But I still got most of them wrong!'

The atmosphere was relaxed and when Peter admitted he'd got them all wrong, Mary suggested there was no hope for the team and the farm.

'But I've got an excuse,' said Peter. 'I've been in hospital.'

'Nice try, Dad,' said Martyn, 'but it was your heart and not your brain that was affected!'

More laughter ensued, and as James explained each answer, they realised that if they'd listened a little more closely, they might have realised what was actually being asked. James explained that most people are not listening but simply waiting to speak – particularly if they don't like what the person is saying.

'But there was one aspect of the quiz that's the most serious lesson to take away,' said James. 'What is it?'

In the silence that followed, James explained it wasn't what he said but 'how he'd said what he'd said' that confused the group. He confirmed that the main reason people leave a farm is not *what* is being said to them, but *how* it's being said.

Knowing nods and smiles broke out around the table as the team thought about examples of people speaking to them in a way that annoyed, upset or confused them. James could see the recognition in their faces as he'd seen it at every workshop he'd delivered.

'So let's find out how we can speak to each other in a way that works rather than a way that doesn't. You'll have heard the phrase, "treat other people in the way you wish to be treated"?

The team nodded in agreement. 'When it comes to manners and respect, this is true, isn't it? Although we do need to be aware of different cultures and practices. But when it comes to communication and the way we *speak* to each other, it's completely wrong. When we communicate with others, we should speak to them in the way *they* wish to be spoken to – and this will make all the difference.'

There were more nods of agreement from the team. 'So let's get stuck into the real reason we're here today. Let's find out how you all prefer to communicate, how you're likely to work together and why knowing your differences could transform your team into a powerhouse of productivity!'

The relaxed atmosphere was now fully established and even Andrew, who'd been sceptical that a team meeting would solve the challenges on the farm, could sense a new mood among the group. There was a togetherness that had been missing for a while and even Charlotte seemed relaxed and ready to listen. Mary was inwardly glowing. Her conversations with the Chairman had taught her to trust her intuition and she knew she'd done just that by booking James Glover.

'I'm going to give each of you a piece of paper with words listed on it,' said James. 'I'll explain what you need to do in a moment.'

There was a buzz of anticipation in the air as the team wondered what was coming next…

14

'Wow! That was amazing,' said Martyn as the team gathered outside in the yard after the session with James. 'At least you all know why, as an Adventurer, I'm always late and get so easily distracted! And perhaps you'll listen to my crazy ideas now?'

Charlotte pulled on her overalls and boots. 'Remember what James said – you can't hide behind your profile, Martyn. Being an Adventurer doesn't mean you can just turn up late for things. It means you must work harder to be on time for those who have more Investigator traits in their profile. Remember, he said we all need to flex our styles and take a step towards each other? Or were you too distracted to remember that part?'

'Very funny, Charlotte,' said Martyn. 'I was just saying at least we understand each other a little more now.'

'I can see why I can irritate you at times, Charlotte, and I'm sorry,' said Amy, also pulling on her overalls. 'I don't mean to be hesitant or indecisive and as James explained, I do worry about getting things wrong.'

Charlotte took a deep breath. 'Actually, Amy, I think I owe you an apology. I understand now that you're a Team Maker, so I know I can probably come across to you as a bit blunt at times. I didn't realise that my Visionary style was going to upset you. I just like to crack on and get things sorted, you know?'

Amy blushed. 'Thanks, Charlotte, but at times I've been a bit hesitant and overly cautious. I'll try to be a little more decisive and take responsibility from now on.'

'Remember, James said that no one needs to change,' added Andrew. 'We all have our different styles but it's our differences, not our similarities, that give our team its strength. We just need to flex our approach with each other in giving or receiving communication and recognising that not everyone sees the world through our eyes. I know that, as an

Investigator, I can be a bit inflexible at times but I've been doing so many things around the farm in the same way for so long, it's hard for me to change.'

Marek was scrolling through his phone and seemed distracted. 'I cannot find the right word in Polish for my profile. Google Translate says an Investigator works for the police. Could you tell me again what Investigator means?'

'It just means you like to know a lot more detail and have a plan, Marek,' said Charlotte. 'You won't like it if people work with you in the parlour in a different way to you. Like when you milk with Martyn and he changes the milking routine: you don't understand why he would do that.'

'I don't change routines!' shouted Martyn.

'You do!' everyone else shouted in unison.

Laughter filled the air as everyone had finally put their overalls and boots on ready to go back to work.

'I actually enjoyed it far more than I thought I would,' said Billy. 'I've not sat in a training room for years and I thought it was going to be boring. But I enjoyed it. I can see why I get on my wife's nerves too!'

There was more laughter from the team.

'OK, everyone,' said Charlotte. 'Let's get back to work. Amy, go and check the calves. Oh sorry, I mean, Amy could you go and check the calves, please?'

Martyn cheered and clapped his hands in a sarcastic manner. 'There you go, sis – it wasn't so difficult to be nice, was it?'

'Sod off, Martyn. I was just taking on board what James said.' Charlotte smiled and inside it felt good. Asking rather than telling Amy wasn't much of a change and Amy confirmed that she preferred to be asked rather than told. If she could make their working relationship better, Charlotte recognised it would certainly reduce her stress.

'No problem, Charlotte. I'll go and check on Dottie too to make sure she's OK and if I need to report anything, I'll call you straight away.'

Observing the conversation, Andrew couldn't believe how simple changes in communication style were already making a difference.

Peter called them to attention from the door of the farmhouse. 'Come on you lot, let's get on, there are cows to be milked, fields to be cut and money to be made!'

He stepped into the yard and James Glover followed behind him. As the team began to disperse, Peter could hear the banter and chatter as the team remarked on each other's profiles – VITA Profiling was the name of the system that James had used.

Peter turned to shake James's hand. 'I think you've made a big difference this morning, James, thank you. When Mary suggested this session, I didn't see how it could help me get the farm running more productively. But now, having understood how we're all different, I can see why Martyn was so frustrated with me. I'm cautious and don't really like change while he thrives on new projects, ideas and opportunities. I think I've stifled him, James. I'm the reason he's leaving the farm. I can see it now.'

James put his hand on Peter's shoulder. 'It's not your fault, Peter. No one is to blame. You have two children who want different things from and for the farm and behave very differently. We do our best as parents, managers and leaders but no one teaches us how to spot these different traits. We all live our lives knowing we aren't all the same but not understanding what our differences mean. That's why we developed the VITA Profiling system. I wanted to stop people from leaving their family farm or any other farm because they couldn't get on with another member of the team. We can't solve every personality clash but we can massively reduce them by understanding a little more about each other. Even you and Mary understand a little more about each other now that you know she's a Team Maker.'

'Oh gosh, she's a Team Maker all right,' said Peter. 'All she wants is for everyone to get along and be happy. She hates conflict and of course blames my unwillingness to embrace change as the cause of most of it – particularly with Martyn. So, what happens now, James – what are the next steps?'

'We'll get all the VITA reports to you within the next 48 hours,' said James. 'You need to share these with the team so they can read each other's reports as this really helps to cement the understanding of their differences. I'm going to use this information to help me in my search for your farm manager as well, as it's important they fit into your team.'

'What sort of person do you think we need then, James?'

'I think they need to have a fair amount of Team Maker in their profile but also need to make key decisions quickly and easily. Visionary traits such as decisiveness and being results focused would be useful too. Ideally, they would also be organised, structured and planned, while being enthusiastic and open to new ideas. But that would make them practically perfect, and perfect people don't really exist do they, Peter?'

Both men laughed. Peter liked James Glover. He seemed to genuinely want to help the team work together more harmoniously while also being more productive and effective. James liked Peter Wilson too. He could see he was a typically honest, hardworking and down-to-earth farmer who was committed to his family, his farm and his staff. Like many other farmers James met, he could see that Peter had grown up on a farm and not ventured much beyond the farm gate to learn about how other types of businesses were run.

He recognised that the farming industry is a fairly closed environment and many of the leadership and management skills found in other industries appeared to be lacking in farming. But James loved it. He'd worked outside the farming industry for many years before starting his own company and when he first encountered farming through a dairy farm in Devon, he couldn't believe how little training was provided to farmers and their teams. Now his company was one of the leading providers of training and development to the agriculture sector in the UK but he still loved to be on a farm, observing people when they understood, often for the first time, why they struggled to work with their family or a colleague. It was rewarding work and seeing Peter and his family have the breakthrough they'd just experienced was one of the reasons James sought to grow REAL Success into a larger organisation that could help more farmers and their families.

'How has the recruitment been going, then?' Peter asked. 'Do you have anyone in mind for our role?'

'It's been going well, Peter. There are a couple of strong candidates who've applied and I'll be speaking to one of them this afternoon. I've already given her details to Mary and hope to get an interview set up by the end of the week.'

'Great news, James. I look forward to seeing the details and CVs soon.'

'You've changed your tune, Mr Wilson,' said Mary, emerging from the farmhouse. 'I dragged him kicking and screaming to this workshop, James, so you've worked miracles already.'

Peter laughed. 'You're always right, dear. You should know by now.'

'Pity you didn't listen to me earlier then, you stubborn old fool.'

And they all laughed together.

After more small talk and discussion about next steps, Peter and Mary watched James drive away from the farm. Mary slipped her arm into Peter's.

'That was quite a morning wasn't it, love?'

'Yes, it certainly was. Thanks for pressing ahead. I can see you were right. And I'm sorry about Martyn. I think we've lost him for a while but hopefully he'll be back one day.'

'Oh yes, our son will be back, don't you worry.'

But Mary *was* worried. She was worried about Peter's health, the family, the team and the farm and it all felt a little overwhelming. But there was another feeling as well. She knew what it was. It was hope. And she felt a nudge in one of the quiet, silent corners of her mind to call and update the Chairman.

15

Mary headed back to the farm office, picked up the phone and dialled the Chairman. Once again, he answered within a couple of rings and she updated him on what had been happening, including the workshop with James Glover. She confirmed that it had created a completely different atmosphere around the farm.

'Sounds like a productive session, Mary,' he said.

'Yes, it was. Discovering our differences and having them explained in such a simple way has really helped the team to understand each other. I'm hoping it will help Charlotte in particular, as she has struggled to accept that people can't always reach her standards.'

'Our job as leaders isn't to try and change people, Mary, but to help them find their own strengths,' he continued. 'One of the biggest lessons leaders need to understand is that our role isn't to make others more like us. It's to help them become the best version of themselves. It's why people often dislike receiving feedback because the leader or manager simply wants their colleague to be more like them. Your people don't want your feedback, Mary; they want your dedicated attention. This means spending time on recognising individual strengths.'

Mary knew instinctively that David was right. Every time she heard Charlotte or Peter telling a member of the team how to improve what they were doing, it was based on wanting that person to carry out a task in the way they would do it themselves.

'And remember,' continued David, 'if people love their work, their work will love them back through improved performance. Just because someone is good at something doesn't mean they actually love it. Finding out what people really enjoy doing is just as important as improving their skills and knowledge.'

'Thank you, David,' said Mary. 'Thanks to today's session with James and your continued wisdom I've realised that to build a great team and truly maximise our strengths and minimise our differences, we need to understand each other. Then, as leaders, we need to build people up using the attributes they already have rather than wishing they were more like us.'

'Exactly, Mary. Good luck with the appointment of your farm manager and go well.'

Mary heard a click on the line so she knew David had ended the call. It was as if he had said all that needed to be said. Mary reached for the notepad where she'd written down key thoughts from her previous calls with David.

Find the silence.
Recruit in haste – regret at your leisure.
Read between the lines.
Trust your intuition.
A breakdown is often a breakthrough.
Watch for the sequences.
Put family first.

She thought about what she wanted to add and wrote down two sentences David had said to her which felt profound.

Our job as leaders isn't to try and change people but to help them find their own strengths.
Your people don't want feedback; they want your dedicated attention.

Mary sat and thought about those two statements and realised that the VITA Profiling session they'd just completed with James Glover was all about helping the team to see who they were rather than who they weren't. It was also about helping them to talk to each other in a way that made them feel they were listening to rather than just speaking at each other. It was clearly a big moment for the team and the family. Mary wanted to make sure the momentum wasn't lost as the memories of the session faded, although the small workbooks given to them during the session would help to keep the discussion alive.

She also wondered about the recruitment of the farm manager. She couldn't recall giving David the details but he seemed to know they were recruiting. 'I wonder what personality style the person needs to be to fit into our team?' she thought. She knew Charlotte would prefer them to possess Visionary traits – to be direct and get to the point quickly, while Peter, Andrew, Billy, and Marek would all like them to have a fair amount of Investigator traits, ensuring they'd work in a planned, organised and systematic way. Amy would want the farm manager to be high on the Team Maker style where kindness, being considerate and generous with their time and keeping everyone calm was important. Martyn wouldn't care who the new farm manager was if they let him try out his new ideas, was friendly, sociable and ready to have fun.

Mary simply wished for someone who was calm, fair and willing to make tough decisions but could also bring the team together rather than create tension and division. Her own traits of Investigator and Team Maker meant she wanted the new person to be clear about what they wanted from everyone and to communicate this in a warm, friendly and supportive way.

The office phone began to ring, interrupting her thoughts.

'Hi Mary, it's James.'

'Crikey, James, are you missing us already? You've only just left!'

Mary could hear the road noise so knew James was calling from his car.

'Yes, Mary, I know – but I've just spoken to Claire Davis and she's really keen to meet you. The team have carried out the screening process and got her to complete a VITA profile, which suggests she's an unusual combination of Visionary and Team Maker traits, meaning she's a straight-talking person who's keen to achieve results through collaboration and positive teamwork. I think she would fit very well into your team, Mary.'

'That sounds encouraging, James. So what happens now?'

'When are you free this week to see her?'

James continued to explain to Mary the process they would go through to ensure the interview went smoothly. He gave clear instructions on how to set up the interview, ensuring Peter was prepared and available to spend the time with Claire.

'She may wish to bring her partner too so they can look around the accommodation you may decide to provide. It might be worth having a member of the team ready to show them both round as part of your interview. We'll carry on looking for other candidates but for now, I think Claire would be a strong addition to your team.'

After a few final pleasantries, they ended the call.

'You look like the cat that got the cream, Mary. What's up with you?' said Peter as he came back into the farm office.

Mary wasn't aware she was smiling. 'I've just got off the phone to James Glover.'

'Blimey, he's keen, isn't he?' said Peter. 'He's only just left us. What did he want this time?'

'He's spoken to Claire Davis and she's keen to come for an interview.'

Peter sighed. 'I'm still not sure about a female farm manager...'

'How can you think and talk like this when your own daughter is working so effectively on our farm?'

'But that's different,' said Peter. 'Charlotte is a family member and the guys in the discussion group all accept their own female members of the family. But to directly employ a female, particularly in a management position, from outside the family isn't something many of them are comfortable with.'

'That's ridiculous,' said Mary. 'You and I both know Charlotte is incredibly capable of making decisions and doing all the physical work around the farm. And to suggest a woman can't manage others is something you and I are going to have to disagree about, Peter.'

Peter looked a little sheepish. 'I'm sorry love, you're right. I guess I spend too much time with farmers of my generation. It's true the younger men in the group don't see it as an issue at all. Anyway, let's not fall out about it. Let's see who this girl is and go from there, yes?'

Peter realised Mary was unhappy with his reaction. He walked over and kissed her on the cheek. 'Sorry, Mary. Don't allow this grumpy fool to dishearten you. You and I have worked together very well over the years and I guess I'm just a little old school in my thinking. Thank you for all you're doing to find us a new farm manager.'

Mary's demeanour softened slightly. She knew her husband's attitude was common throughout the agricultural sector. 'I know, Peter. But James Glover thinks Claire would fit really well into our team and having just done a profiling session, we must bear this in mind and look for the right personality rather than the right gender.'

'Yes Mary, you're right. Just ignore this fool and crack on. I'm going to make a couple of calls to suppliers as we need some feed.'

'It's about time you tidied up your desk' said Mary, still feeling slightly miffed with her husband.

'Yes but I know where everything is, Mary. I'll tidy up eventually.'

'James was saying that we need to create the right environment for the interview with Claire and I don't want her thinking we run an untidy farm when she comes into an untidy office,' said Mary. 'We need to give the best impression, which means being ready, prepared and on time, as well as showing her our genuine appreciation for coming to see us.'

'Yes, Mary. I'll sort out the desk this morning.'

'Just see that you do, Mr Wilson,' said Mary, leaving the office and heading back into the house.

16

After her call with James Glover, Claire started to feel excited. The way he'd explained the challenges at the farm, the variety of work and the characters in the team made her feel that this could be the opportunity she'd been looking for. She'd become frustrated in her current role, primarily due to the difficult working relationship with the farm owner who was at times so blunt that he came across as aggressive.

The farm where she currently worked was part of a large estate but despite her employer being well educated and from a family of farmers going back generations, he was incapable of managing people in an effective way. Wilson Farm sounded like an ideal environment for her to reignite her passion for farming, feel more appreciated and work with a dedicated team.

James Glover had asked her to complete an online questionnaire and when she received the VITA profile report about her personality style, she was surprised by how accurately it described her. She often found herself confused by her own desires and ways of working as she always wanted to get along with others, didn't really like conflict and believed in the power of the team. But she would become frustrated when results didn't come fast enough or people were indecisive. She was ambitious and wanted to achieve her ambitions as quickly as possible. This meant she often came across to others as a little impatient and would find herself apologising afterwards for being demanding. James said her profile contained both Visionary and Team Maker traits which, while being an unusual combination, was potentially a strong leadership profile as it combined a desire for achievement with effective collaboration.

James also supplied an overview of the team and promised to send her individual VITA reports so that Claire could get a deeper understanding of everyone ahead of her interview. James also talked about Peter and Mary, who sounded quite different to her existing employer.

Claire was sitting in the farm office having just finished her call with James when Edward Longstaff, the farm owner, burst through the door.

'What the hell is going on this morning, Claire? I've just seen two of the lads in the parlour spraying each other with water rather than washing down properly. What do you think this is, a circus? I don't pay staff to lark about and waste my water in such a stupid way. And I don't pay you to sit in this office while that's going on. Go and sort it, please.'

Claire didn't even bother to respond. Edward made no effort even to say 'Good morning'. She knew Freddie and Michael had a tendency to indulge in horseplay around the parlour but their intentions were good. Both had struggled at school and were often accused of not concentrating and being distracted in class.

From the way James was describing Charlotte's brother, Martyn, Freddie and Michael sounded like him. James referred to Martyn as an Adventurer who was keen to have fun and bring light-heartedness, even to serious situations. Claire left Edward without saying another word and headed over to the parlour. By the time she got there Freddie and Michael were just finishing off the washdown.

'Morning, boys,' said Claire. 'How's it been going this morning?'

'Good thanks, Claire,' said Freddie. 'Michael was being an idiot and chucking water at me but I got him back by chucking dried-up slurry at him, which hit him in the face. It was hilarious. Then Edward came in and blew his top. I thought he was going to explode!'

'Yes,' said Michael. 'I've never seen him so angry. But I suppose we should be used to it by now. He never has a positive word to say to any of us. He just criticises, moans and then shouts at us.'

'Yes, I know he can be difficult,' said Claire, trying to remain professional. 'But what impression do you think it gives others if you're larking about and not getting the job done?'

'But we *do* get the job done,' said Freddie. 'Tell me, Claire, does this parlour look clean to you?'

Claire looked around the parlour. It was spotless. 'Yes, it does, so well done, but just remember, not everybody thinks having fun is part of the job. We've got to get things done too. Anyway, get yourself cleaned up, go and have your breakfast and be back for morning routines. OK?'

Michael and Freddie both gave Claire a thumbs-up and carried on hosing down the parlour. As Claire walked away, Freddie turned to Michael, waving at him to turn off the hose.

Claire was exhausted by the constant peacemaking and having to pick up the pieces after Edward's latest outburst. She recalled her last conversation with James Glover and he'd suggested the Wilson family needed help – and she liked to help. She also enjoyed seeing positive results and there seemed to be more that could be achieved at Wilson Farm. It seemed like a perfect combination and in direct contrast to her current role.

She had worked hard to improve the working conditions on the farm but even when she requested a new kettle for the staff room, Edward complained and wondered why the old kettle – just like the tatty sofa and the broken fridge – was anything other than 'perfectly adequate' for his staff. She'd ended up buying a new kettle with her own money simply because she couldn't face Edward's reaction to her requests.

She'd also suggested several improvements to the way the cows were managed. And yet, if it involved any form of spending, Edward was unwilling to invest in the fabric of his farm, which looked tired and tatty. When people arrived for interviews and had to negotiate the farm drive with its deep potholes, overgrown hedges and piles of rusty machinery, she knew she'd struggle to convince potential employees that this would be a good place to work. She'd often spent her own weekends and evenings with a hedge trimmer trying to make the entrance of the farm look more appealing. Many potential employees were positive by the time she'd finished talking to them but then Edward would show them around the farm. On their return, they seemed far less enthusiastic. It was an exhausting treadmill Claire had decided she needed to step off.

Claire decided to call Annabel. She knew what Annabel would say as she'd spent many hours complaining to her about the farm. Annabel had listened patiently and been as supportive as she could.

'Is everything OK?' asked Annabel nervously. It was unusual for Claire to call during working hours.

'Yes, everything's fine; I just wanted to let you know I've got an interview. You've been telling me for ages that I needed to find the right job and two things have come together to make me realise I'm ready to move on – Edward's latest outburst and a call from James Glover at REAL Success.'

'Alleluia!' screamed Annabel. 'Maybe now you'll stop moaning so much!'

'I know, I'm sorry. You know I hate to admit defeat but I don't think I can change Edward. Freddie and Michael were messing about a little bit at the end of milking this morning and yes, while they should've been a bit more focused, there was no harm done. Crikey, if Edward got covered in cow slurry every morning, along with the monotony of having to wash down the parlour every day, I suspect he might get bored too. But he went berserk at them and was rude to me too. I just don't need it, Annabel.'

'About time,' Annabel replied. 'You've been a saint on that farm and he's had this coming for a while. Where's the interview?'

'It's a family farm in south Devon that needs a farm manager. It's a mixed farm and would be a step up from what I'm doing now. I've been on farms with sheep and arable and to be honest, I quite like the thought of the variety. I think I'm ready for this now and I don't really want to spend another day sorting out Edward's disasters.'

'Sounds like you've already made your mind up.'

'Yes, I have. It could be so exciting. Thanks for listening as always.'

'No problem. Catch you later.'

Claire sighed deeply. Annabel was such a good listener. She never judged and often didn't even express a view. She was just prepared to listen to Claire's ramblings and offer her support. It was a priceless gift and one she treasured, having known Annabel for most of her adult life. As she walked back to her house at the edge of the farm, she felt energised and wondered about the synchronicity of Edward's outburst preceded by the call from James Glover. This had happened before in her life, when events seemed to link up in perfect harmony at just the right time. She wasn't religious in any way but found herself looking up towards the sky. 'Thank you,' she said.

17

Charlotte was struggling to hold it together. She'd felt like this a few times recently but now the feeling was worse than ever. There was a tightness in her chest and she felt like she couldn't breathe. She was frightened too. She wasn't sure what she was frightened of, but she felt scared, as if something awful was going to happen. These uncomfortable feelings had broken her sleep a few times recently and now she was sitting on the edge of her bed wanting to crawl back under the covers but knowing she needed to get up as it was her turn to milk. She looked at the clock and it said 4.15 am. Her hands were shaking and she felt lost and confused. She took a few deep breaths to try to calm herself down but the tightness in her chest and feeling of foreboding remained.

She felt weak and incapable of making any decisions and didn't think she could face the team. Thoughts were tumbling towards her and she couldn't stop them. 'Come on, Charlotte,' she said to herself. 'You can do this.' She slowly got up and headed towards the door. As she placed her hand on the handle, she considered turning round and getting back into her safe and inviting bed. She couldn't face another day of feeling pressurised and anxious.

She hadn't told anybody how she was feeling as she was worried about appearing weak and unable to handle the pressure of the farm. Today was an important day as Claire Davis was coming for her interview and Charlotte wanted to give the best impression of herself and the farm. And yet, her bed called to her as a safe haven from the stress and pressure beyond her bedroom door. 'Come on Charlotte, you can do this,' she said to herself once more. She went to the bathroom and looked into the mirror, barely recognising the person looking back at her. 'Gosh, you look awful, mate. Time to pull yourself together.' She took three deep breaths and felt herself calming down a little.

Within a few minutes she was ready to leave the house and headed for the milking parlour. The lights were already on and she could see the cows in the collecting yard. Martyn and Billy were moving the cows towards the parlour.

'Morning, sis!' called Martyn. 'What time do you call this?'

Martyn was rarely ready for work before Charlotte and when she didn't respond, he called out again. 'Hey, sis, everything OK?'

Charlotte still didn't respond and kept walking towards the parlour. Martyn was puzzled. Something wasn't right.

'Billy, I'm heading into the parlour to help Charlotte. Keep pushing the cows up, please.'

He walked through the narrow gap in the collecting yard where he found Charlotte in the parlour checking the teat spray gun. He walked up and slapped her on the shoulder.

'What's up? Not like you to be cutting it so fine when you're milking!'

Charlotte turned towards Martyn. 'I'm fine, thank you. Just didn't sleep too well last night.'

'You look knackered, mate. Are you sure you're OK?'

'I've told you, I'm fine. Let's just get on, shall we?'

Martyn wasn't convinced. His sister didn't look well and it looked as if she'd been crying.

'Come on, sis, you look awful. Are you sure everything's OK?'

'Today is an important day, bro,' said Charlotte. 'We've got this interview with Claire and I've got to trim feet this morning, plus the vet is coming later too, so it's going to be a busy day. Just have lots on my mind.'

Martyn wasn't convinced by his sister's response but he knew she was stubborn so there was no point in pushing his questioning any further.

'OK, sis. But if you want to talk about how you're feeling, just come and find me. We're all feeling the pressure and I know I've not helped by taking the job in Cornwall, so I'm sorry if it's added to your stress levels.'

Charlotte didn't respond because inside she was still angry with her brother for leaving the farm. She was also sad that they wouldn't be working together anymore, even though he frustrated her at times. They milked the cows without saying much to each other. There was the occasional chat about a cow but the milking went smoothly.

When the last cow had left the parlour, Martyn started the washdown and Charlotte went into the parlour office to update a few records. She'd noticed during milking that she'd started to feel calmer and the pain in her chest had disappeared. She'd focused her thoughts and energy on the cows and it seemed to stabilise her nervous system.

Then Peter appeared at the office entrance. 'Morning, Charlotte. Is everything OK? Martyn said you were a bit quiet this morning.'

'I'm fine, thanks, Dad. Just didn't sleep very well last night. I have lots on my mind now with Martyn leaving and Claire coming today.'

'It's my job to overthink things, not yours!' said Peter. 'You don't need to get stressed about anything, Charlotte. My heart attack has made me realise we need help and I think you need help too. This is a big farm and for too long we've tried to keep costs down and do a lot of the work ourselves. But it's a false economy. I got ill and to be honest, you don't look well this morning. So we have to change something, don't we?'

The discomfort in Charlotte's chest started to return and worry and fear crept into her mind again. 'I'm fine, Dad, honestly. I'm looking forward to meeting Claire today.'

Charlotte placed her hand on her dad's shoulder, leant forward and kissed him on the cheek in a rare show of affection. 'Love you, Dad.'

'Blimey. You must be feeling strange if you're giving your old Dad a kiss at this hour of the morning!'

Peter watched Charlotte walk through the parlour and head back to the house. He was worried about his family. Charlotte looked tired and drawn, Martyn was leaving and Mary was talking to people outside the farm about their difficulties. He wondered where he'd gone wrong. He recalled what Benjamin the hospital chaplain had said to him about difficult situations becoming the start of something positive but all he'd ever wanted was for the farm to become a successful and safe place for his family to live and work. But his son didn't want to work with him anymore and now Charlotte looked exhausted. 'Gosh, I hope this Claire Davis is good today,' he thought. 'We clearly need somebody to help us.' He walked back towards the main farm office to find Mary busy tidying up the papers on his desk. Peter hadn't tidied up as he'd promised so Mary had decided to intervene.

'Hey, I told you to leave those papers where they were!'

'If I waited for you to do anything round here, dear husband, nothing would ever get done. Claire is due to arrive in about half an hour and I wanted the farm office to give the impression that we're organised and everything is calm. Anyway, I've put things into sensible piles of invoices and letters and I'm throwing most of these magazines away because you never get to read them. You may as well cancel the subscription because what's the point in paying for a magazine you never read? There must be at least a year's worth of *Farmers Weekly* and *Farmers Guardian* magazines still in their wrappers here.'

Peter picked up a few of the magazines from the pile. 'I should read them, as there's often important information in there about what's happening in the industry. But I can never find the time.'

'That's why we're interviewing Claire, isn't it? Hopefully having a farm manager will give us all a little bit more space to breathe and to sit and read a magazine from time to time. That would be amazing wouldn't it, love?'

'Yes dear, it would. So am I showing Claire round the farm, or should we do it together?'

'Neither, I'm afraid. James suggested that Charlotte should show her round after we've done the more formal interview.'

James Glover had sent Mary a detailed framework for the interview, which she'd looked at the previous evening. He insisted that they must complete the formal part of the interview first. He said it was far easier and more professional to move from the formal interview to the informal farm tour.

'He sent across a list of interview questions we can ask Claire and suggested we make notes while she's talking so we can discuss it with him after the interview. I've printed out the questions here.'

'Do we need to do the serious interview bit?' asked Peter. 'In the past, we've just chatted while showing people round the farm, which seemed to work OK.'

'James said it's important to show a professional image because Claire is a professional herself. We often forget that people like her who are building their careers treat their jobs and where they work very seriously. We haven't really given an interview the consideration we should have.'

Peter sat down on one of the chairs, which he now looked at with different eyes. They were clearly old and tatty and he realised he'd never considered that candidates were appraising them too.

'I can see where James is coming from and to be honest, no one has ever taught me how to do an interview. I've just done what I thought was right and what others in my discussion group have said they do. It seems there are quite a lot of things I've done wrong, though. It feels a little bit like it's all falling apart in front of me.'

'Don't be silly,' said Mary. 'This farm has been here for a long time and is going to be here for a long time after we've gone. Yes, there are things we need to do differently to help the farm work for us but the world has changed and people have changed. We just need to up our game a little. It starts today with Claire's interview. I'm determined to give the best impression we can of the farm. Which reminds me, have you fixed the farm sign at the front of the drive yet? It was hanging off the other day, which doesn't give a good first impression. In fact, I noticed there are rusty old bits of metalwork on the right-hand side of the farm drive too, which should've been disposed of a while ago. Plus there are a few old tractor tyres stacked up against one of the gates. Do you think we should try and tidy those away before she gets here?'

'I'm not going to have time,' said Peter grumpily. 'It isn't the King visiting, you know. It's just a potential new employee.'

'Well maybe this is our problem,' said Mary. 'Perhaps if we considered the King might be coming to visit, the farm would look tidier than it does. You could at least just fix the sign before Claire gets here. From what James said to me, she's likely to be early too.'

Peter got up and left the farm office without saying another word. As he walked towards the workshop to pick up his tools, doubts began to return about whether employing a farm manager was a good idea. He didn't want or need anyone walking round his farm criticising all the things he hadn't done properly. He wanted help, not criticism. As he passed the two tractor tyres and the old bits of machinery and metal he just never got round to moving, he began to realise Mary was right. It did look as if they didn't care. By the time he'd reached the end of the drive he'd also noticed all the potholes and puddles, a gate held together with string and bits of plastic floating around the field. He'd never really looked at the farm through the eyes of a visitor before. It was a sobering moment and when he saw the farm sign hanging off the post, held on by just one screw, he felt it was a metaphor for the whole team. Everybody was clinging on by a thread and needed support and maintenance. Charlotte certainly looked that way.

He carefully reattached the farm sign to the post. Even the sign was a little faded and he made a mental note to replace it. It made him think of his own position on the farm and how it was perhaps time for him to step back and be replaced by somebody younger with new ideas and fresh energy. As he walked back down the drive, he spotted guttering hanging off the corner of one of the sheds and more gates that needed fixing. Rather than feeling down and depressed, he felt a strange sense of excitement and anticipation that somebody might make suggestions about how the farm could be improved. By the time he'd taken the tools back to the workshop and returned to the farmhouse, he was looking forward to meeting Claire. His mindset had shifted and his mood had lifted.

Mary put fresh cups and biscuits in the middle of the kitchen table and was busy wiping down the kitchen surfaces and cleaning away any clutter.

'I don't think Claire is going to decide whether to work for us on the basis of how tidy our kitchen is, Mary!' said Peter.

'I know,' said Mary curtly. 'But first impressions count and as we're going to bring her in here first before we go to the farm office.'

'Yes, you're probably right,' said Peter. 'I hadn't thought about that. In fact, as I came down the drive, I noticed a few parts of the farm looking tired and tatty. While it doesn't make a massive difference to how the farm performs, it could make a big difference to what people think of us when they arrive. I've fixed the farm sign now but I've decided I'm going to replace it with a more modern-looking sign. I'm also excited to hear what Claire has to say when we take her round the farm. A fresh pair of eyes is exactly what we need right now, Mary.'

Mary stopped cleaning and put her hands on her hips. 'That's one of the most sensible things you said for quite a while, Peter Wilson. If Claire is the right person for us and we are the right farm for her, today could be the beginning of a new chapter. Let's pull out all the stops, shall we? I've already spoken to Charlotte, Martyn and the rest of the team to ensure everybody is aware that Claire is coming and we need to give the best impression we can. I'm quite excited too. Let's just hope Claire doesn't have any awkward surprises in her background, eh?'

Peter looked out of the kitchen window towards his farm. He felt a pang of emotion in his chest when he thought about his dad and his grandad, who had no doubt stood in the very same spot many times, wondering about the future of the farm. He knew there would have been times when they were concerned about progress and he recalled conversations he'd had with his father about new ideas. He remembered his father saying, not long before he died, that the farm needed to keep moving forward and that Peter was just a steward of the farm. His job was to protect it, manage the resources carefully and then pass it on.

'Are you OK, love?' asked Mary.

'I've just realised that today could be the first step in me doing what Dad asked of me: to hand the farm over in a better state than I received it.'

Mary stood next to Peter and held his hand. 'Gramps would be so proud of you.'

'Do you think so, Mary? Even with me having a heart attack, Martyn leaving and Charlotte looking exhausted? Have I made the right choices for my family?'

'You've always done your best and that's all we can do both as parents and on the farm. It's up to Charlotte and Martyn to decide if they want to carry on running the farm but maybe having a farm manager will make the decision easier for them, not harder. Let's see how today goes, shall we? But always remember, your children love you, respect you and want you to be happy too.'

Peter turned to Mary and kissed her on the cheek. 'I love you, Mary Wilson.'

'You too, love. But now it's time to roll out the red carpet as Queen Claire is arriving!'

As they held hands, they watched a car heading down the farm drive. And both Mary and Peter wondered what the future would bring for all of them.

18

Claire was nervous and excited. She'd found the farm easily and the entrance was clearly signposted. As she navigated the farm drive, she noticed a few bits of metalwork lying around and a couple of the sheds in the distance looking a little tired, but overall her first impressions were positive. Her Ford Focus battled with the potholes in the drive but this wasn't unusual and no worse than Edward's farm. By the time she pulled in front of the farmhouse, two people were emerging from the front door. She assumed they were Mary and Peter Wilson. Claire parked the car, took a deep breath and opened her door.

'Hello, Claire, it's lovely to meet you.'

Claire shook Mary's hand. 'Lovely to meet you too, Mrs Wilson.'

'Oh gosh, call me Mary, please. This is Peter, my husband, overworked farmer and my hero.'

Peter laughed and put out his hand. 'I'm not sure about the hero bit but the other parts are true. Nice to meet you, Claire. How was your journey?'

Mary peeked inside Claire's car. It was spotless inside and Claire was dressed in a crisp blue shirt, dark blue jeans and black boots. Her brown hair was pulled back into a ponytail and Mary guessed she was about 5 ft 8 in tall. Her first impression was positive and she hoped Claire felt the same about them.

Mary ushered Claire into the farm kitchen and offered her a seat at the table. 'Let's have a cuppa first and then we can move into the office for the interview.'

Peter smiled as he watched his wife do what she did best – make people feel welcome. They were soon chatting away about their drink preferences and eating biscuits. Claire asked about the family and Peter gave her a little history of the farm and talked about Charlotte and Martyn. He didn't mention Martyn was leaving.

'James mentioned you have a partner, Claire, but no children?'

'Peter!' Mary scolded. 'You're not supposed to ask that! Particularly as Claire has only just arrived!'

'Don't worry, Mary,' said Claire. I do have a partner but no children as yet, although we'd like to have them one day.'

'What does your partner do?' asked Peter.

'She works as a vet. She's a good one, too. All her farmers love her.'

'She?' thought Peter. '*She*?'

'What's her name, Claire?' asked Mary, trying to act casual while wondering what Peter was thinking.

'Annabel. Although I call her Belle. We've known each other for most of our adult lives. We went to the same school and kept in touch when we went to college. We've lived together now for four years. We'd like to get married soon.'

'Married?' Peter was still reeling from the news that Claire was in a relationship with a woman and now she was talking about getting married. He was stunned into silence.

'How exciting,' said Mary, trying to keep the atmosphere light while knowing that her husband, who'd gone very quiet, was in shock.

'Yes, it is. We've talked about having a family too. It all seems a bit early to be talking about this with you and I wouldn't want my plans for a family to scupper my chances of taking this role. I'm fully committed to developing my career and so is Belle. We would plan things carefully, so please don't worry about it. It's just that we were talking about your family, who are obviously important to you, and mine are to me too. I have two brothers and I'm the youngest. My mum and dad have been supportive of me and Belle, although Dad took a bit of time to get his head around it.

'I'm not surprised,' said Peter without thinking. 'Sorry, I mean you just want your children to be happy, don't you?'

Mary gave Peter a stern glance to get him to stop talking. He was in danger of offending their guest and potential new employee.

'I'm very happy thanks, Peter,' said Claire, familiar with many people's reactions to her choice of partner. She was fully aware that farming remained an extremely male-dominated industry and that sexual preference wasn't usually a topic of conversation. While she'd met several gay people within the industry, they'd found it hard to be fully accepted. Claire, however, was ready to answer any questions thrown at her so Peter's reaction didn't faze her at all.

'We're delighted you're happy, Claire,' said Mary. 'Annabel would be very welcome to come to the farm to check out the accommodation we're going to provide as you'll both need to feel it's the right move for you, particularly if you intend to start a family here. Shall we move over to the office so we can start the formal interview? Then we can show you round the farm.'

'Sure, let's get started,' said Claire.

Peter was still thinking about Claire and Annabel being a couple on the farm. He was wondering how the team would react, as well as the local community and his farming friends. Would they approve or disapprove? It was confusing and not the start of the interview process he expected. He'd imagined a man with a young family coming to set up a new life. But with Claire there were no children and uncertainty as to how she and Annabel would fit into the team. It all felt very unsettling.

Peter had a glazed expression and Mary knew he was playing scenarios over in his mind.

'Come on, Peter, we're moving into the office now,' said Mary.

'Sure, yes, sorry, was just thinking about something on the farm… I'm coming.'

Once inside the farm office, they all sat down on the chairs that Mary had carefully arranged.

'OK then, Claire,' said Mary. 'We'd like to ask you a few pre-prepared questions and then give you the chance to ask us any questions you like about the farm, our family or our plans. This is your time too, so please don't consider any question too silly or insignificant – we want you to be as certain as we are if we both decide to move forward.'

James Glover had briefed Mary closely on how to open the interview and while Mary hadn't discussed it with Peter, he seemed happy to let her take the lead.

'So, tell us a little about yourself and your career to date, Claire.'

Claire outlined why she'd chosen farming as her career, which was partly due to Annabel's love of animals and her veterinary career. Claire had helped out a few times at the practice and visited farms with Annabel. They all chatted in a relaxed manner about Claire's career and her reasons for considering leaving her existing role.

'To be honest, Claire,' said Peter, 'I'm often criticised for being too soft or slow to make decisions. I want everyone to be happy at work while still ensuring we have high standards and follow protocols. I don't believe shouting at people gets a better result, unless you're cheering on your rugby team, of course!'

They discussed Claire's experience and understanding of livestock issues and examined her knowledge of arable and sheep farming, which was certainly less developed than her knowledge on the dairy side of the business. However, Peter reassured her there were experienced people in the team who handled those areas and they'd help her get up to speed.

When they invited Claire to ask questions, she asked about how Charlotte and Martyn felt about a new person coming into the team above them when they would eventually become the owners of the farm. Peter explained Martyn's decision to work away from the farm for a while to gain some different experience while Charlotte was committed to the cows but was less interested in the sheep and arable aspects of the business.

Then they discussed salary, working hours and rotas. Claire was straightforward but respectful – just as James predicted she would be. When Mary had read Claire's VITA Profiling report, she understood why James described her as being a little unusual. She liked to speak directly and to the point but also cared deeply about those around her. This came through strongly when Claire asked searching questions about farm performance (which Mary and Peter couldn't answer) while also expressing interest and concern about animal and staff welfare.

They discussed her notice period and outlined the details of the property they were proposing would be made available. It was a small cottage on the edge of the farm which was rented privately by a couple from the village, but they'd given in their notice. Claire remarked on the coincidence of the house being available and suitable for a small family, before the formal part of the interview concluded.

'Charlotte will show you around the farm now, Claire,' said Peter. 'I'll just her call to say you're ready. She should have finished milking by now but I'll check.'

Peter stepped out of the office, leaving Claire and Mary alone together.

'Do you think Peter is OK with me having a female partner, Mary? He seemed quite shocked when I told him about Annabel.'

Mary was surprised that Claire asked this as soon as Peter

had left the room but wasn't surprised she'd asked the question.

'Listen, Claire, Peter and I need help. It's too much for him and too much for me as well. We need more time off and we need your experience, manner and personality. While I'm sure you're used to a level of prejudice in this industry, I would want you and Annabel to feel entirely welcome here and to bring up a family if you choose to do so. Peter might be a little old fashioned but he's a caring and loving father. All he wants is for the farm to run smoothly and be here for Charlotte, Martyn and their children if they wish to continue to farm.'

Peter came back to the farm office with Charlotte in tow.

'Hi Claire, I'm Charlotte, how are you?'

'Hi Charlotte, I'm great, thank you. I've been finding out all about the farm so I'm excited to see it now.'

Charlotte and Claire headed towards the Kubota parked in the yard, leaving Peter and Mary behind.

'So what do you think then, Peter?' asked Mary.

'I think she's lovely.'

Peter and Mary continued to discuss Claire's skills, experience and suitability for the farm. They concluded Claire was indeed exactly what they were looking for. They just needed Charlotte to be positive and upbeat about the farm.

'Charlotte looked very stressed this morning when I saw her in the parlour. I'm a bit worried about her to be honest.'

'Don't worry. I suspect Claire and Charlotte are going to get on.'

'Let's hope so, love,' said Peter. 'Let's hope so.'

19

'So how was your interview with Mum and Dad?' asked Charlotte tentatively.

Claire was also feeling nervous, concerned about what Charlotte was thinking. She knew it was important for Charlotte and Martyn to feel comfortable with her.

'It was a really positive conversation,' said Claire. 'Peter and Mary seem like nice people.'

'Most of the time!' laughed Charlotte.

Claire laughed too and that released some of the tension between them. They both clambered into the Kubota and Charlotte began to describe their surroundings.

'I'm not sure how much Mum and Dad told you about the farm but we have just under 800 acres of land, mainly used for pasture for the cows and to grow food for them. We sell a little on to local farms but it's mainly grown for our girls.'

As they drove out of the yard and turned down the lane they chatted about the farm and Claire asked lots of questions, which Charlotte answered openly and in detail. Charlotte was equally impressed with Claire's knowledge of the arable side along with the flock of sheep they saw as they drove across fields and lanes before returning to where the cows were grazing. Charlotte suggested they got out to have a look at the cows. They discussed the farm's approach to rotational grazing and walked towards the cows, which were happily munching their way through fresh pasture. The herd was a mix of Friesian and Holstein and Claire was impressed with their condition.

'The cows look well, Charlotte, and happy too. You must have been working hard to get them into this condition.'

'It's been quite hard to be honest, Claire,' said Charlotte. 'I feel as if I'm the only one who really cares about them and I get frustrated with members of staff when they don't notice if the cows are ill. We nearly lost one of my favourite cows the other week because Amy simply forgot to check in on her when she was a bit sick.'

'That must be so frustrating,' said Claire. 'I'm passionate about the welfare of the animals too so maybe I'll be able to offer you a bit of support to make sure the team are as focused as you are in keeping the animals well.'

Charlotte felt a mixture of excitement and relief. Finally, somebody else would be able to help carry the load she'd been under for a while.

'Have you felt under pressure at times?' asked Claire.

'I have felt the pressure a little bit in the past few weeks,' said Charlotte.

'I know how it feels,' said Claire. 'I've suffered periods of anxiety over the past couple of years, working for a man I've struggled to understand. I often wondered whether I was doing a good job for him.'

'Gosh, how awful,' said Charlotte. 'I think my pressure is mainly what I put on myself more than anything Mum, Dad or anybody else puts on me. I just want to do a good job and for the farm to be successful, the cows to be well and productive. I often feel it's all my responsibility even though I know it isn't. Does that make sense?'

'It makes total sense,' said Claire.

Just as they were about to climb back in the Kubota, Claire stopped and placed her hand on Charlotte's arm. 'It's important to me that if I get the job on your farm, you know and feel I'm here to support you and not take your place. I understand the pressure you're feeling and would really like to help. Feeling anxious can often translate into physical symptoms too, like a tight chest or feelings of worry or fear. Again, I know, because I've felt it too. I even paid for help from a therapist on how to control panic attacks a few years ago when I really felt I couldn't cope. Have you ever felt like that?'

There was a pause. Charlotte didn't know Claire and yet felt instinctively she could trust her. She hadn't spoken to anybody about the way she'd been feeling and yet Claire already understood both the pressure she was under and the physical symptoms she'd been hiding.

'The problem is that everyone thinks you've got to be strong, work hard and never give up. This whole industry is built around hard work, long hours and showing no sign of vulnerability. So I've not wanted to admit I'm struggling. But to be honest, I am.'

Charlotte started to feel emotions she'd suppressed now welling up inside her and tears heading straight towards her eyes.

'I understand,' said Claire. 'This is a crazy industry dominated by men who are afraid to talk about their emotions and a culture of long hours, hard work and never showing any sign of weakness. It's ridiculous. We need to change this. In fact, we need to change this industry to one where people can talk about how they're feeling rather than worrying about showing any vulnerability. Too many people suffer in silence and never express how they truly feel. I have female and male friends in the industry who feel like you and end up leaving the industry. It's so sad but I want you to know, if I end up getting the job, I'm here for you. I know we've only just met but I will do everything I can to help you not only to be the best farmer you can but also the happiest version of yourself.'

The emotions finally took over. Charlotte's shoulders dropped and the tears began to flow. All the worries, fears and feelings of inadequacy she'd been holding onto for months finally found their way out. Without saying a word, Claire stepped forward and put her arms around Charlotte, holding her tightly.

No more words were spoken but against the backdrop of the animals Charlotte loved, in the fields where she worked, on the farm she hoped to inherit one day, she sobbed uncontrollably in the arms of somebody she'd only just met. Charlotte couldn't see, but Claire was weeping too. She felt Charlotte's pain deeply and knew, even more than after her interview with Peter and Mary, this was the farm where she needed to be. The two women bonded in their understanding of each other and quickly established trust. They stood in silence for a few minutes more. As the emotions subsided and the tears slowed, it was Charlotte who spoke first.

'Sorry, Claire, I don't know where all that came from.'

'Ah, it's OK. I've never cried at an interview before either, so I guess we're both on new ground!'

They both laughed and the tension eased even more but the bond between them was firmly secured. Charlotte knew she wanted to work with Claire. She was going to convince Peter, Mary and Martyn or anybody else who asked that Claire was the right person to be there for her in the future.

They clambered into the Kubota and headed to the milking parlour. They chatted in a relaxed manner about the equipment, milking times and routines, as well as discussing interests outside the farm. Claire told Charlotte about Annabel. Charlotte knew a friend at college who was gay and she'd kept in touch since they'd left. She no longer worked in the agricultural sector, partly because she never felt truly comfortable being who she was. Claire's honesty and openness around her sexuality simply made Charlotte even more keen to have Claire working on the farm and leading the team. It made no difference who Claire loved. All that mattered was that she cared for her cows in the same way as Charlotte did.

Martyn and George were also in the yard, cleaning down farm machinery. Charlotte hadn't managed to find Andrew and Billy as they were in the far reaches of the farm looking after the sheep. As they headed back towards the farmhouse, Mary waved through the window. She got up quickly to come out and talk to them.

'What do you think of the farm, Claire?'

Claire was surprised at how she felt about the question. It was her turn to feel emotional. She knew this was the place she wanted to work but didn't want to look so keen that she sounded desperate. She took a deep breath and paused before replying. Mary was worried that Claire was about to be polite but not positive.

'I think the farm is amazing, Mary. I don't want to sound too pushy but I really think I could make a difference here and I'd enjoy supporting you all too. Charlotte has done an incredible job and just needs a little bit more support, especially with Martyn leaving. I hope we can show Martyn that this is where his long-term future is and entice him back.'

Mary almost leapt around the yard with joy. She'd known the moment she handed Claire over to Charlotte that this was the person she wanted to manage their farm. Mary was on edge the whole time Charlotte was showing Claire round the farm in case her daughter might be a little blunt or too direct.

'I'm delighted you think so, Claire. Let's head back to see Peter and then you can be on your way. And we can talk about next steps.'

'It was great to meet you, Claire,' said Charlotte. 'I hope we see you again soon. Really soon!'

Charlotte smiled at Claire and then instinctively reached out and gave her a hug. Mary was taken aback at Charlotte's display of affection towards Claire, whom she'd only known for a short time. It was unusual for her to express her emotions in this way and Mary wondered what they'd been discussing.

'Great to meet you too, Charlotte, and remember, you're doing an amazing job but when things get tricky, as they always will, just talk to someone, OK?'

'I will. Thanks again, Claire.'

Charlotte wandered back towards the parlour and Claire turned to Mary.

'Your daughter is a strong woman and we need more strong women in our industry, don't we?'

'We certainly do. Let's go and have a cup of tea and talk about what happens next. Peter is in the office waiting for us.'

'Hello, Claire, so what do you think of our little farm, then?' asked Peter.

'First of all, I'm not sure it's a little farm, Peter. It's slightly bigger than I imagined with all the land you've got, but I think, and I know you'd expect me to say this in an interview, that the farm is amazing. The cows are in great condition too, which must be down to the brilliant team led by Charlotte, who is doing a great job, I might add!'

'Thank you, Claire, we've all worked hard, too hard perhaps, to keep the cows in good condition.'

'Yes, Charlotte is working hard,' said Claire. 'We talked about the importance of speaking up when things get too difficult. I've been there myself over the years as I'm sure you both have too. One of the challenges in this industry is that vulnerability is often seen as a sign of weakness.'

'You can say that again,' said Mary. 'I've been saying to Peter for years that the men in this industry drive themselves into an early grave far too often. Any talk of retirement for men in their sixties and seventies seems like a death threat to them! All they've ever done is work tirelessly on their farms, day in day out, and then they wonder why families break up and people leave the farm. It's all so sad and unnecessary.'

'I'd like to change that, Mary,' said Claire. 'We can work fewer hours if we have the right number of people on the farm working the right sort of rotas. We can all have a life outside the farm if we ensure people have enough time off to follow hobbies and interests. And I don't think it's a sign of weakness to admit when you're struggling either.'

'Well, Claire, I think you're right,' said Peter. 'For far too long, all I ever thought about was the farm. The sense of responsibility of running a family farm and handing it on to future generations is a huge pressure for me and many of the people in the industry and particularly the men I speak to. You're so right that we need to rebalance this but for many of us we just don't know where to start! I was never really taught by my dad how to manage people, or even the basics of HR. Mary and I have just done what we thought was best. But we need the next generation now, and people like you, Charlotte and Martyn to help us do things differently.'

'Well, you can start by giving me the job!' said Claire with a huge smile on her face. And they all laughed.

'Let's talk about next steps,' said Mary. 'I think we're all in agreement that you should come down with Annabel, maybe over a weekend, and work on the farm for a couple of days, just so you can get a feel for the farm and for us. Then you can look round the local area and make sure it meets your needs, particularly as you're thinking of starting a family. There are great schools locally where both Charlotte and Martyn have fared well, so we can show you those and give you all the details.'

'Gosh, I think we might be getting ahead of ourselves talking about schools and I'm not sure what Annabel would be saying if she knew we were already talking about starting a family!'

They all laughed again. Peter and Mary continued to feel an instinctive sense that Claire was a good fit with their family. James Glover had clearly selected the right person and even though they hadn't interviewed anybody else, they both knew Claire was the one they wanted to join their team.

Peter and Mary watched Claire get into her car and disappear up the drive.

'I think we've found a new farm manager, Mary,' said Peter.

'I think we've found more than that,' Mary replied. 'I think we've just found the answer to our prayers.'

20

Martyn had briefly met Claire the previous day and liked her. He could see she was going to help his family and the team. However, he had a sense of 'missing out' and not being on the farm to see the improvements Claire might bring. He also felt she would have supported the ideas that his dad had so far dismissed. However, today he was going to the new farm for a couple of days before he started properly the following week. Determined not to be late, he'd set two alarms and was showered, dressed and ready for his trip to Cornwall with sufficient time to make a coffee before starting his journey.

As he entered the kitchen, Charlotte was already there making herself a drink before she started work.

'Morning,' said Martyn.

Charlotte looked up to acknowledge him.

'How are you feeling about going to the farm today?'

'Well, it's interesting, sis. If you'd asked me a couple of days ago, I would have said that I was genuinely excited. I was thinking about the prospect of working for a more relaxed boss and driving the new John Deere too. But after meeting Claire yesterday, I've now got mixed feelings about it all.'

'Why?' asked Charlotte.

'I'm not exactly sure,' said Martyn. 'I guess I only met Claire briefly but I liked her at once and can see how much she's going to help the farm if you all decide to give her the job. I'm curious to know how it's going to work out and am gutted I won't be here to see it.'

'But that's your choice!' said Charlotte. 'You decided you wanted to work somewhere else and leave us in the lurch, so I guess you'll have to just hear from us how things are going while you're down in Cornwall.'

Charlotte still struggled to mask her frustration with Martyn's decision. She felt he was being selfish and it could harm the farm. She also struggled to hide her own disappointment and sadness that her brother didn't want to work with her anymore.

'Yes, I guess you're right, sis, but I've committed to this now and I've got to see it through. I can't suddenly change my mind because of Claire. And I do want to see what it's like working on another farm.'

Martyn made his coffee and sat down at the table opposite Charlotte.

'So how are *you* feeling about Claire coming to work here? Do you think she will stop you from developing your career and taking over the farm?'

'Nah. I don't think so. She would be a brilliant addition to our team and has more experience than me so I can lean on her, which would be cool. I've felt for a while, and I'm sorry brother but this is true, that I've been carrying the farm and now I need support. You're not going to provide it, so I'll be delighted to have Claire come in and help me.'

'I'm sorry if you think I haven't been pulling my weight. You and I are very different and I guess we tackle things in different ways. You worry about things more than I do whereas my view is that things will always work out in the end. I guess I'm just a bit more of an optimist than you are.'

Despite sounding so certain, Martyn was covering up how he truly felt. He was worried about his sister and had noticed her demeanour getting more and more downbeat over the past few weeks, particularly since their dad's heart attack. It wasn't like Charlotte to be depressed. She'd always been the strong one of the two and talked of plans and visions for the farm, but recently this had slowed down and completely stopped.

'How are you really feeling, though?' asked Martyn, trying to get his sister to open up a little.

'I'm fine. I'll just be grateful for the support and I think Claire could provide it. It'll be good to have another woman on the farm too.'

Martyn decided to try once more. 'Yes, I'm sure it will be good to have another female in the team. But how are you truly feeling? Come on, sis. Be honest. We might not have many more chances to talk as brother and sister. So while we're here in our family kitchen where we've grown up together, I want you to talk honestly with me.'

Charlotte felt the irony of this moment. She could always talk to her brother. When they were younger, they would share everything that was going on at school and in their lives but more recently, when the farm became their primary focus, they'd spent less and less time together. And now she wasn't sure whether to tell Martyn how she'd been feeling or of the conversation with Claire the previous day. She took a deep breath.

'I've found the past few weeks a bit tough and at times I've not felt that well. It's been affecting my sleep and occasionally I've felt a bit panicky and worried about whether I can cope.'

Charlotte stopped speaking as she could feel the emotions rising once more. Martyn reached across the table and held his sister's hand.

'It's OK, sis. If I'm honest, you completely amaze me. You're like an engine in a tractor that never cuts out. You work and work and work and don't really take time off, so I'm not surprised your body has started to complain.'

The tears started to edge out of Charlotte's eyes and she wasn't sure whether she was going to be able to hold them back. There was so much she wanted to say to Martyn. She wanted to beg him to stay on the farm to support her and yet she also wanted him to make his own life choices.

'I talked to Claire about this yesterday and she told me about periods in her life when she felt anxious, scared and overwhelmed. It was helpful to talk about it and she gave me a few tips. I also spoke to the Farming Community Network and they signposted me to other agencies I can speak to if I experience a similar feeling again.'

Martyn looked across the table at the sister he admired so much. The feeling that he was deserting the family overcame him and he felt regret about his decision to leave the farm. He continued to hold Charlotte's hand.

'I'm sorry, Charlotte. I'm so sorry if I've been selfish in wanting to move away from the farm. But you know how frustrated I've been. I still feel I've got to do this but I know it's putting more pressure on you and I'm sorry. You know you can always talk to me and I wish you'd spoken to me before about the way you were feeling.'

Charlotte looked at her brother. 'Yes, but you'd made up your mind about what you wanted to do so what was the point in me talking to you about how I'm feeling about the farm? With Dad's heart attack, I genuinely thought it would be you and me having to run the farm, particularly if we lost Dad.' Her voice broke and trailed off.

'I know,' said Martyn. 'When Dad was lying in hospital and I was waiting for him to wake up, those thoughts came into my mind too. Maybe you and I would have to run the farm and I'd have to change my plans. But Dad pulled through, didn't he? Anyway, I think it's been a wake-up call for him. Maybe if you do employ Claire, he'll listen to her. He wasn't prepared to listen to me. I want to come back to the farm one day but I need to try something new. I've become bored with the daily routine on the farm and I need something different. You know what I'm like. I get bored easily, so I'm hoping this will give me a new lease of life and then perhaps, in a few years' time, I can come back to the farm and help you once more.'

'Yes, I know. While we might be siblings, we need to live our own lives and I want you to be happy. I've known for a while you were frustrated with Dad's reluctance to change things on the farm and I've shared similar frustrations. I understand why you want to leave, so go off to Cornwall today and have a good day with Barry and his shiny new tractor.'

Charlotte squeezed Martyn's hand and then let go. Martyn took a few gulps of coffee and then rose from the table.

'I will,' he said. 'The Wilsons are a tough bunch and I believe in you, my big, brave sister. You'll be fine. If Mum and Dad ask about my opinion of Claire, just tell them I thought she was great and I think you should employ her.'

'I will,' said Charlotte. 'Now go away and let me finish my coffee.'

Martyn came round the table and kissed his sister on the cheek.

'I love you,' he said.

'Sod off, you soppy git,' said Charlotte.

As Martyn walked across the yard, the mixed emotions he'd felt when he'd woken up were still there. But then he remembered the brand new tractor at Burton Farm and how it felt when he sat in the cab. He began to get excited once more about the prospect of driving the tractor and being a key part of Barry's team. Martyn drove up the farm drive past the farm sign and noticed it was no longer hanging off its fixings.

The journey down to Burton Farm was uneventful and quick. He watched the sun rise and had plenty of time to think about his family, his sister, his life and the potential future he was walking into. When he drove down their farm drive, a trickle of nerves surfaced again. By the time he arrived at the yard Martyn had another feeling he couldn't quite place. It felt like a gentle sense of foreboding or a warning. He dismissed it, cut the engine and got out of the car to be greeted by Barry.

'Morning, Martyn!' said Barry, with a great big smile on his face. He shook Martyn's hand and slapped him on the back. 'Great to see you and welcome back to Burton Farm. Today is about you getting to know the team, the farm and our girls before you start in a couple of weeks' time. So just shadow me for the day and I'll show you round. I'm not sure who's milking this morning but we'll go and check out how they're getting on first. I've got a bit of work to do with the cows before I can show you the new machinery I've invested in. I should be able to let you go late afternoon so you can get home at a sensible time. Does that sound like a plan?'

It sounded like a vague plan to Martyn but he was excited about the new machinery and was looking forward to seeing it.

'Come on then,' said Barry. 'Let's go and see how the guys are getting on in the parlour. It's a beautiful morning, isn't it?'

Martyn loved Barry's optimism and positive outlook and could feel his spirits rising. The nagging doubts when he arrived were beginning to dissipate. He wanted to be around somebody who was innovative, positive and optimistic and Barry seemed to represent what Martyn had been looking for. The excitement he'd felt on his last visit returned as they walked over to the parlour and chatted enthusiastically. By the time they'd reached the parlour, Martyn was certain he was embarking on a new and exciting future.

21

Mary woke up in a good mood. The previous evening, Peter and Mary had talked at length about the farm, the children and their future. Both felt Claire should be a part of their plans. They'd reminisced about their journey at the farm so far and realised how quickly time had passed.

Peter had already left the farmhouse and was out working so Mary showered, dressed and headed down to the kitchen to make herself breakfast. Even though it was early, she decided to give James Glover a call. As she reached for her mobile phone, she felt a tinge of nervousness and excitement. This call felt like the first step into an exciting future and she was suddenly frightened Claire might not accept what she was going to propose to James.

'Good morning, Mary; how are you this morning?' said James.

'Hi James, sorry to call you so early.'

'That's fine. I'm already working anyway. Got a busy day ahead today with training on another farm, so I was just getting my bag ready before I left. So how did it go with Claire yesterday?'

Mary felt giddy with excitement but wanted to sound professional and calm.

'Well, James, the interview went very well. We followed your format and used your interview questions and it worked so much better than other interviews we've done in the past. Peter was a bit nervous about the more rigid structure but I can see how it worked and hopefully, gave a more professional impression to Claire.'

'Claire called me as she was leaving the farm,' said James. 'She's incredibly excited about the prospect of working with you. So whatever you did, it worked! I've rarely talked to a candidate who's so enthusiastic about a role. She said it felt like destiny that she'd arrived not only on your farm but on that specific day. She said she connected very well with Charlotte too.'

'That's so good to hear, James. Peter and I discussed our situation last night, the package we could offer Claire and her suitability for our business. We're convinced she's the right person to take our farm forwards. So what happens now?'

James explained the next steps in the process, including the importance of a professional offer in the form of an email, which he would organise. They discussed the details of salary, start date and the accommodation and James took notes to ensure he presented the offer in the right way.

'How about a start date? Did you discuss this too?'

'Yes, we did,' said Mary. 'Claire said her contract required her to give her employer three months' notice.'

'Well, her existing employer is a bit unpredictable,' remarked James. 'While I haven't recruited for him directly, I've heard about him through the industry. Several farmers have said to me he seems to get through staff quite quickly and Claire has been one of the longest-serving members of the team. I wouldn't be surprised if he asks her to leave earlier than her three months' notice.'

They discussed the final elements of the offer and James promised to get back to Mary when he'd spoken to Claire. Mary sat for a few moments wondering whether Claire would accept their offer. 'Of course she will.' The voice in her head surprised her but it felt like her own intuition reassuring her all was well. Mary decided to ring David. She had no real intention for the call but sensed a nudge to call the Chairman. He answered within a few rings.

'Good morning, Mary. What a delight to hear from you. How are things progressing on the farm? Have you made any important decisions yet?'

Mary wondered whether James had spoken to David. It was always as if he knew what was going on even though she hadn't spoken to him.

'Yes, David. We've decided to take on a farm manager. We interviewed her yesterday and Peter and I have both agreed to offer her the position.'

'Well done, Mary. How do you feel about your decision?'

David simply listened without interrupting as Mary talked through her concerns and excitement.

'Do you think we're doing the right thing?' asked Mary.

'That's not for me to say, Mary. Your intuition will tell you if you're doing the right thing. Does it feel right?'

'It does, David. For the first time in a long time, I sense we're moving forward. At times, I was unsure if we could get through each day without collapsing in a heap. With Peter's heart scare, we knew something needed to change.'

'It's easy to ignore the nudges though, isn't it?' said David. 'But those nudges can become a push, followed by a kick, followed by a resounding thump until we listen to what our intuition is trying to tell us!'

'Yes, you're right. As always, I appreciate your time and wisdom, David. Thank you for your support with our decision about Claire.'

'Go well, Mary.' The phone clicked and David was gone.

Mary reached for her notepad. She turned the page and re-read the notes she'd already taken from her conversations with the Chairman.

Find the silence.
Recruit in haste – regret at your leisure.
Read between the lines.
Trust your intuition.
A breakdown is often a breakthrough.
Watch for the sequences.
Put family first.
Our job as leaders isn't to try and change people but to help them find their own strengths.
Your people don't want feedback; they want your dedicated attention.

She decided to add three words:

Follow the nudges.

Mary sat in the office for a few more moments, noticing her breathing and the silence around her. A sense of calm descended and she closed her eyes. 'All is well,' said the voice inside her head. As she looked through the window towards the farm, Mary felt a sense of peace she hadn't experienced in a long time. It wasn't only the fact that help was coming. It wasn't just the voice in her head either. It was the connection to her intuition that felt so reassuring. For so long, she'd been saying to Peter that they needed help and he'd bluntly refused to even consider help until he had his heart attack. Mary sensed she'd been nudged to make change happen and eventually the kick, as David described, arrived in the form of Peter's health warning.

Mary wondered if God existed. She wondered about David, who he was, and how he seemed to know so much about their lives. While she hadn't set foot in church for a long time and didn't follow any meaningful faith, she found herself looking up to the sky. All she said was, 'Thank you.'

22

Peter was already missing Martyn. He'd only been gone a couple of days but his absence was noticeable. Even though they'd regularly disagreed, not having his son around just didn't seem right to Peter. It gave him a heavy feeling in his chest. He'd decided to see Andrew and check in on the sheep, so he headed over to the far side of the farm in the Land Rover. As he drove the short distance, memories stirred of times he and Martyn were in the Land Rover together. From an early age, Martyn would help him with whatever tasks he was planning on doing that day. As Martyn got older and eventually started to drive, it would often be Peter in the passenger seat as they headed off towards the outer edges of the farm or to move some cows. As Peter drove down the narrow lane towards the furthest point of his farm, he longed for those days and realised how quickly they'd passed.

Then he felt a pang of guilt because he knew his daughter was still very active and present in the business. He loved working with Charlotte too and they rarely shared a cross word as they had very similar views on how to run the farm, despite the fact that Charlotte was often more frustrated with the staff and certainly more blunt.

Peter wondered what Martyn was doing and visualised him on the new tractor he'd spoken about so enthusiastically. He wondered whether he'd been too cautious as the tractors were showing their age and needed constant repairs and maintenance. Perhaps if he'd loosened the reins slightly and allowed Martyn to choose a new tractor, he might still be working on the farm? He turned the Land Rover into the field where he could see Andrew herding the sheep towards another gate, clearly planning on moving them into a new field.

The sun was shining and it was a dry, crisp morning. Peter reflected that it was the sort of morning when you were glad to be a farmer. It compensated for the poorer weather when, through the darkness and the cold, you'd wonder whether farming was as enjoyable as you'd hoped it would be. But today was a good day and as Peter walked towards Andrew his spirits began to lift.

'Morning boss,' said Andrew. 'How are you, sir?'

'It's a cracking morning, isn't it, Andrew?'

Peter's disappointment about Martyn's absence from the farm was already beginning to fade and as he met up with his most loyal employee, he wondered why Andrew had stayed on the farm for so long.

'The ewes are looking good,' said Peter.

'Yes, they are. It won't be long now. Lambing is still my favourite time of the year, despite the fact it's a bit manic and the days can be long. But it's sort of why we do what we do, isn't it?'

'It sure is,' said Peter. 'Can I ask you a question, Andrew?'

'Sure, shoot.'

While they were talking, Andrew whistled commands to his sheepdog Bertie.

'Why have you stayed on our farm for so long?'

'That's a big question for this time in the morning but easy to answer. It has suited me, that's why. The job always felt right as I love sheep, the land is beautiful, you and Mary have always looked after me with little drama or crisis and I've simply loved being a shepherd. I guess it's always ticked all of my boxes. Never really thought of looking elsewhere, to be honest. Can I ask you a question in return?'

'I guess so,' said Peter, slightly nervous about what Andrew was about to ask.

'Why did you ask me that question? Is it because Martyn has left the farm?'

'Yes, I suppose so. I'm already missing him. Even though we were arguing a fair bit before he left, now that he's gone, I can already feel the hole he's left behind. I'm just wondering why he couldn't stay, particularly as he's my son, while you have stayed for so many years.'

'I think there's a different dynamic between family members, Peter. The way a father or mother speaks to their children, or vice versa, can be worse than the way they speak to their staff. Martyn is very different to you. He's a bit of a hothead, has tons of ideas and gets bored easily. So perhaps he was just ready for a change or the idea of working on the same farm for many years didn't appeal to him. I think the next generation can feel trapped by their family farm and Martyn just wants to spread his wings for a while. He'll be back, mark my words.'

'I hope you're right because I think I may have pushed him away,' said Peter.

As they moved the sheep from one field into the next with Bertie's help, they chatted more about the farm and its history. Andrew talked about Peter's dad and how he and Peter were similar, unlike Peter and his son Martyn. Peter found it comforting to speak to Andrew, who had become a dear friend over the years.

By the time they'd finished gathering the sheep, Peter was already feeling much more relaxed. When he returned to the Land Rover and glanced at the empty seat next to him, he simply wished nothing but success and happiness for his son. He hoped he was having a good day driving the new tractor and was going to be happy on his new farm. He turned the Land Rover around and headed back towards the farm. As he drove back along the lanes and fields he knew so well, he thought about his father and how he'd run the farm so effectively before him. They'd rarely argued, mostly agreed and always enjoyed each other's company. He missed his dad, who always seemed to know the right thing to do. It had been comforting to have his father there to talk to and now, with Martyn gone from the farm as well, Peter felt his absence so deeply. 'I miss you, Dad, and I hope I'm making you proud,' thought Peter. 'Every day,' replied a voice inside his head.

A wave of emotion came over him and he thought his mind was playing tricks as the voice in his head seemed to come from the empty seat next to him. 'Wish you were still here, Dad,' he said to the empty space. 'I'm always here,' was the response in Peter's head. As he drove back up the lane towards the farm, he wiped tears from his eyes. He couldn't remember the last time he'd wept, as he preferred not to show his emotions. He felt immensely proud of his son, who'd been brave enough to leave the comfort and certainty of the family farm to develop himself in an unknown and unfamiliar place.

He was also proud of his daughter and his wife and how they played their part in running the farm. As he thought of his family, the tears gently flowed and he felt a wave of pride and gratitude. After a few moments, the tears subsided and he wiped his eyes on the back of his sleeve. As he drove back into the yard, he could see Charlotte heading towards the house for her breakfast. Peter parked the Land Rover and glanced into the rear-view mirror to make sure it wasn't obvious he'd been crying. He took a deep breath and got out of the vehicle.

'Morning, Dad, are you OK?'

'I'm more than OK, Charlotte. I'm so proud of you, Martyn and your mum. In fact, I'm proud of all of us. We have an amazing farm, family and team. What more could I wish for?'

'Blimey, Dad, where did that lot come from?'

'I'm not sure to be honest but it's true, Charlotte. I'm so proud of you.'

Charlotte linked her arm with her dad's.

'I love you, Dad.'

'You too.'

They walked towards the farmhouse arm in arm, anticipating a well-earned breakfast.

23

Claire was looking forward to speaking to James Glover. She hadn't slept too well but realised it was simply due to excitement. She'd talked with Annabel for hours and they'd looked at all the different aspects of the move to Wilson Farm and she couldn't find any reason not to take the job if it was offered to her. Annabel was convinced she could find a new job near the farm and although they needed to look at the house on offer, based on what they knew, it sounded just as nice as where they were now.

It was Claire's day off so they planned to drive to Wilson Farm and explore the local area. It was unusual for them to have a day off together, and Annabel remarked how fortunate it was that this day fell just a few days after Claire's interview.

'So how are you feeling about the job this morning?' asked Annabel.

Claire wasn't sure how to answer. She was a jumble of nervousness, excitement, trepidation – but above all, joy.

'I don't know where to start, really. It just feels so right. When I've looked for jobs before, there's always been a compromise I needed to make. But for this job, I can't find any necessary compromises at all. There's a part of me wondering if I'm suppressing something or not noticing an aspect I should be thinking about. Do you think I'm looking at it through rose-tinted spectacles?'

'I don't think so,' said Annabel. 'If it feels right, it probably is right.'

'OK. I'll speak to James today to ensure that before I accept any offer, I've got all the information I need to make the right decision. I'm also going to thank him for matching me with a farm and a family I know I'll enjoy being part of, and where you and I can be happy.'

Annabel looked at Claire and could see a sparkle in her eyes she'd not seen for a long time. They'd been together for seven years and Claire had worked on two farms, moving house each time. With Claire's last job she'd seen the sparkle disappear completely. She'd often come home and cry about another incident. They would talk things through, Annabel would build Claire back up, and Claire would go again the following day. Claire's resilience never ceased to amaze Annabel. She was a fighter and didn't like giving up. But now, life was changing and Annabel was excited for them both.

'I'm so thrilled there's a sparkle back in your eyes,' said Annabel. 'You know I'll follow you wherever you go but all I want is for us to happy and for you to be happy at work. I know I'll find a role, so don't worry about me. If we're together and enjoying what we do, our lives outside of work will be so much better too.'

Annabel served up the scrambled egg on toast she'd been preparing and they sat and chatted about their future. They hadn't talked about long-term plans as it was hard to plan when life felt so uncertain.

'Why don't you call James Glover now and see if he has an offer from the Wilsons?'

Claire's heart was filled with a mixture of nerves and excitement. 'Doesn't that seem a bit desperate?'

'Not at all,' said Annabel. 'Go on, call him now.'

Claire dialled James's number and he picked up within a few rings.

'Hi Claire, great to hear from you. How are you?'

'Well, it depends on whether the Wilsons want to make me an offer.'

'Yes they do, Claire. Following our quick catch-up yesterday, I spoke with Peter and Mary and they'd like to offer you the role. I will confirm this in an email to you today.'

'Fantastic news, James, and I'd be delighted to accept the offer if it's based around our discussions on the farm.'

They discussed Claire's three-month notice period and the fact that her existing boss might allow her to leave early. However, if he insisted she remained in her post, Claire wanted to honour her contract. James confirmed that he would let Peter and Mary know.

'Well, I guess there's no turning back now,' she said to Annabel after the call ended. 'They're offering me the job.'

'That's brilliant news, Claire; I'm so pleased for you. Let's get ourselves ready and we can head down to Devon to look around. Do you think it's worth letting Peter and Mary know we're coming in case we could look at the house?'

'Once James has confirmed he's spoken to them, I'll ask whether he thinks it's the right thing to do. But I think it's a good idea and I'm sure they'll be fine with it.'

James called back a few minutes later to say how delighted Mary and Peter were that Claire had accepted their verbal offer. James confirmed it was OK for her to ask Peter or Mary about the house and gave her Mary's mobile number.

'Shall I ring Mary now?' said Claire.

'Go for it,' said Annabel. 'This is so exciting!'

Claire dialled the number James had provided and Mary answered within a couple of rings.

'Claire! Thanks for ringing. James has told us you've accepted our offer. We're absolutely thrilled you're going to be joining us.'

Mary invited Claire and Annabel to come to the farm for a cup of tea and look at the property they'd offered them. Even though they'd only met once, Claire felt she'd known Mary for a while and they'd already established a genuine respect for each other.

Annabel and Claire cleared the kitchen, prepared themselves for their day trip and were soon in the car heading south to a new life together.

24

Claire was dreading telling Edward but she knew the next chapter of her life couldn't begin without the last one having closed. After a lovely day with Annabel driving around the area close to Wilson Farm and having the opportunity to view the beautiful cottage the Wilsons were providing, Claire was more certain than ever that they were making the right choice. But she still faced the hurdle of informing her employer of her intention to leave.

When they'd returned from Wilson Farm, she'd texted Edward and arranged to see him after morning milking the following day. She was ready and waiting in the farm office at the agreed time, but Edward was nowhere to be seen. After waiting for 15 minutes, she called his number. It took several rings before Edward picked up the call.

'Yes, Claire?'

'Hi Edward, I was just checking if you were coming in for our meeting at 10 am as we agreed yesterday.'

'Yes I am, and I'll be there soon. Just finishing off a few bits first. I'll be there when I'm ready, OK?'

Claire sat and waited for another five minutes, which didn't help her nerves. Eventually, Edward appeared in the office looking sweaty and flustered.

'I've given the lads a rollicking again this morning, Claire. You really do need to sort them out or else I'll have to fire them myself.'

'I'm afraid you may have to do just that, Edward, as the reason I wanted to see you was to hand in my notice.'

'What? You can't do that! Why? What am I supposed to do now? You should have spoken to me first! That's dropped me right in it, that has. Thanks for nothing, Claire.'

Claire was determined to maintain her composure as Edward continued to provide a long list of reasons why she couldn't resign. He said it was going to leave him in a mess with the staff and he simply wasn't prepared to make any more allowances for poor performance.

'I'm really sorry, Edward but I'll be leaving as per my contractual notice in three months' time, less a little holiday I have left to take.'

'To be honest, Claire, I'd prefer you to leave earlier. If your heart isn't in the job, you're no use to me. Leave in a month once we've sorted out a few bits and pieces. I'm not interested in having staff on the farm who don't want to be here.'

'Most of your staff don't want to be here,' thought Claire.

'OK, Edward. I'll work out how much holiday I'm owed and then we can agree a final leaving date.'

'Where are you going anyway? Have they offered you more money? Everyone leaves for more money and expects us farm owners to pick up the pieces. Drives me barmy.'

Edward's sweaty and angry demeanour showed no sign of abating, so Claire decided to bring the conversation to a close.

'I'm going to be farm manager at a family-run, mixed farm in Devon.'

'Farm manager? I don't think you're ready for that, Claire. You'd have to learn a lot more before I'd make you farm manager here.'

'That's exactly why I'm leaving,' thought Claire.

'I'm delighted they see the potential in me,' she continued. 'Thank you for my time here, Edward, and I wish you every success in the future.'

Claire got up to walk away from the agitated Edward and towards a brighter future.

'Hang on, Claire. You've not said exactly where you're going. I want to ring them and tell them you're not ready to be a farm manager.'

'With the deepest respect, Edward, I don't need to tell you where I'm going and I don't intend to offer your name as a reference. I've found my time here extremely stressful and I'm afraid you'll continue to have staffing issues while you see staff as your problem. Most of the team here find you difficult to work for as you constantly criticise and never praise.'

Claire turned and left the office. As she walked towards the parlour, she felt a surge of pride. While she preferred to be direct, she'd never been brave enough to tell Edward how she really felt. Having just done so, she felt liberated.

Edward, however, was in shock. No one had spoken to him so directly before. Claire was by far his most successful appointment and while he'd never told her that, he genuinely appreciated what she'd done for his farm. He had no idea how he was going to replace her or how to be more supportive of the staff. His father had been tough on the staff and Edward knew of no other way to manage people, having never worked on another farm or in another business. But Claire said he needed to change his approach. She'd always appeared to have an excellent relationship with the staff, who respected her and seemed to want to work for her. He, on the other hand, could never get anyone to do what he wanted and the staff didn't seem to respect him at all. 'Maybe I need to be more like Claire,' he thought. He picked up his phone and called Derek from KLK Services. They'd supplied people in the past and saved him the hassle of looking for staff himself.

'Hi Derek, Edward Johnson here. I've got a problem. My herd manager has just dropped me in it and resigned. Can you find someone to fill the gap?'

'Hello, Edward. Of course. I have several herd managers on our books. I'll make a few calls. Leave it with me.'

'Excellent. Thanks Derek. I'm not too bothered about what they've done or who they are. I just need them on the farm pronto, OK?'

Edward's unsuccessful approach to recruitment and staff retention continued. But it was no longer Claire's problem…

PART THREE

The New Era Begins

25

The next month was a time of planning and preparation for everyone at Wilson Farm. When Claire accepted the position of farm manager and confirmed she could start within a month, the news quickly spread across the team, leading to discussions and debate as to the changes Claire might introduce.

Amy hoped Claire would ease the pressure on everyone else. Andrew hoped Claire would be able to mould Charlotte into a more rounded and competent people manager. Billy could see the benefits of bringing in the right person to lead the team. Marek, who had returned from Poland earlier than expected, was happy that someone was going to organise rotas, jobs and tasks and hoped she'd continue to allow him to return to Poland on a regular basis to help his bereaved family.

Martyn was repeatedly reporting back about how well things were going in his new job. Peter continued to miss Martyn's presence but was adjusting to his son not being on the farm. He thought about him every day and while his first few calls were positive, he'd also mentioned that the mobile home he was living in was a little draughty and damp and Peter worried about Martyn's health, as he'd suffered from mild asthma as a child. But if he was enjoying his work, Peter was able to convince himself it had still been the right decision for Martyn to move to Burton Farm.

Mary was the busiest of them all during Claire's notice period. She organised top to bottom cleaning for the house where Claire and Annabel were going to live. Two rooms were redecorated and all the kitchen cupboards were emptied, cleaned and disinfected. Carpets were shampooed and the garden was given a makeover. The previous tenants hadn't taken care of the garden and Claire had confirmed how much Annabel enjoyed gardening. Mary felt it was important that Annabel settled in quickly, particularly given that she'd need to find a new job.

Ever since her interview, Charlotte had been looking forward to Claire's first day. She'd taken time to talk to her mum and dad, just as Claire suggested, about the anxiety she'd been experiencing. She also confirmed she was getting help from a friend who was a mental wellbeing therapist. Mary was upset that Charlotte was unwell but hadn't spoken to her or Peter about it. Charlotte reassured her parents that Claire also understood how she'd been feeling and it was simply the pressure of trying to step up and take the pressure off her dad.

Peter blamed himself for Charlotte's difficulties and checked regularly to see how she was coping. Charlotte told him to stop fussing but he felt a heavy responsibility for how she was feeling. He was looking forward to Claire arriving on the farm, which would ease the pressure on him and in turn ease the pressure on Charlotte.

Mary and Peter spent many evenings talking through their vision for the farm. Peter insisted that if he were to step back, standards shouldn't drop, and he was a little concerned that Claire might want to spend money on new equipment, implement new processes and change protocols that were well established. Mary suggested that Peter needed to be more flexible and he would have to empower the farm manager to make their own decisions.

Peter had spoken to several of his farming friends who'd begun the process of stepping back and handing over the farm, either to their family members or to a senior employee such as Claire. Tom Bridge, a respected and well-established farmer just four miles away, reassured Peter that bringing in a farm manager from outside the family would be the best decision he'd ever made. He explained how his children were enthusiastic but not ready to take on the full responsibility and floundered a little under the pressure of expectation and lack of knowledge and skills to do the job.

Everyone at Wilson Farm was feeling positive about the arrival of Claire and Annabel. During their visits to the farm, they gradually moved furniture into the house, giving everyone the opportunity to get to know them before Claire started her new job.

James Glover kept Mary informed throughout the process leading up to Claire's first day. He'd encouraged her to develop an induction plan for Claire's first few weeks on the farm and helped her develop a staff handbook as well as a contract for Claire.

James also encouraged Charlotte to put together a series of the key protocols such as milking routines, cow health and other aspects of how the farm was run so that Claire could quickly get up to speed. He also shared with Claire, as promised, the VITA profiles of the team so she was able to understand a little more of each of the personalities before she started to work on the farm.

As the weeks passed by, the whole farm was gearing up for a fresh start. There was just one member of the team who was feeling adrift and couldn't share in the enthusiasm…

26

Martyn hadn't told his mum and dad or indeed anybody else that he was regretting his decision to move to Burton Farm. After just a few weeks, it was clear that the farm was run in a chaotic and haphazard manner, with Barry rarely being present on the farm, preferring to leave everyone to get on with the jobs they thought needed to be done.

Milking routines were irregular and Martyn noticed several of the cows didn't look well. Some were clearly suffering with mobility issues and when he asked Barry about lameness, he simply said that a foot trimmer came once a month. When Martyn expressed deeper concerns that several animals clearly needed help and support, his comments were dismissed and he was told not to worry.

However, several cows appeared to have mastitis and he was surprised by how little the team knew about the cause and treatment of the condition. Despite being treated with antibiotics, a cow had died and Martyn was disturbed by how unconcerned the team appeared to be about the loss of one of their animals. When Martyn confronted Jim, the herd manager, who seemed to spend more time on his phone looking at Facebook than managing the team, he'd been told it wasn't Martyn's concern, and he should just focus on getting the cows milked.

At one point a cow collapsed in the parlour and Martyn was asked to drag her out using the telehandler. The team had laughed and joked about the cow and even kicked and punched her to try to get her to move. When Martyn intervened and told them to stop, he was told, 'This is how we do things around here.'

The novelty of the new tractor had quickly worn off. While it was full of exciting technology, Martyn hadn't had a chance to drive it and had become frustrated with other pieces of equipment that were worn out, often needing adjustments to fit onto the new tractor. He'd spent a long time trying to fit the mower on and it was clear that this wasn't considered when the new tractor was bought.

The biggest issue troubling Martyn was the mobile home in which he was expected to live. At the interview, he'd been given the impression he'd be living in a house and yet here he was in a cold and damp mobile home. Electric heaters barely gave out enough heat at night to keep the home warm and the shower leaked, so every morning he'd have to mop the bathroom floor.

Martyn hoped his working days would be shorter but they turned out to be longer. He was often milking at 4.30 am and still on the farm at 7 pm as the chaotic routines and poor rota management meant many jobs simply took longer than necessary. He was homesick too. He missed his sister (even though she occasionally annoyed him) and the banter with Andrew, Billy, Marek and Amy. He'd sit, night after night, alone in his mobile home and began to realise the draw of a shiny new tractor had turned out to be a complete distraction from what he truly wanted, which was to work on a farm where the animals were well looked after, people worked as a team and he felt valued and appreciated.

Barry had barely spoken to him since he'd started other than to shout across the yard to ask him if he was OK, only to move on and not engage in any form of conversation. Jim, the herd manager, clearly liked to give orders. He constantly referred to the team as 'the lads' even though Jenny, the calf rearer, was clearly not a lad. Jim was from a bygone era where the only men who worked on the farm were those he saw as subordinates and less important than him.

Martyn yearned to be back at his home farm but knew he couldn't ring his mum and dad to tell them because he'd be admitting he made a hasty and rash decision to move. When he'd spoken to Charlotte, she told him about the excitement surrounding Claire's arrival. When he told her about what he'd seen on Burton Farm and how it was being run, Charlotte had encouraged Martyn to come home to enjoy the new era at Wilson Farm but, stubbornly, he refused to accept the invitation and was determined to make his new job work.

27

'Morning, love,' said Peter as he entered the kitchen. 'You're up early this morning.'

'Morning,' said Mary. 'It's an important day, isn't it?'

'I guess so,' Peter replied. He sat down opposite Mary at the kitchen table where so many conversations had taken place over the years.

'I'm not sure you can say "guess so",' said Mary. 'It's one of the most important days in the history of our farm and I'm excited about Claire coming to help us. I was reading an article this morning that James Glover wrote for *Farmers Weekly* and I've torn it out for you to read. James talks about the first 100 days on a farm being the most important ones, so we've got to make them count.'

'Blimey,' said Peter. 'There's no stopping you this morning, is there?'

Mary was slightly frustrated by Peter's laid-back attitude. They'd put so much work into getting to this point that she didn't want the momentum to stop the moment Claire arrived. Once Peter knew Claire was coming, it seemed to Mary that he felt the recruitment process was over. But James Glover had made it clear that the recruitment process didn't end when a candidate arrived on the farm. He said the hard work was just beginning and it was important to put as much effort into Claire's first few weeks and months as they put into the recruitment process.

'I just don't want this to fail,' said Mary.

'There's no reason why it should fail,' said Peter. 'The team have all been well briefed, seem to like Claire and Annabel, and even Charlotte is excited about today. I saw her heading towards the parlour this morning for milking and she seemed upbeat.'

The plan was for Mary to cook breakfast for everyone and they'd combine the introduction to Claire with a brief staff meeting, which was not a common occurrence at Wilson Farm. Peter never felt the need to gather the team together as he'd usually just let people know what was happening individually. But he agreed that a staff meeting to welcome Claire was a good idea. They also had a plan for her first few weeks on the farm, including running through all the financial data, introducing Claire to suppliers, vets and consultants, along with spending time with each of the team members. They had all the paperwork ready to go through with Claire, including the legal checks for her right to work in the UK, along with her contract of employment and the staff handbook that James had been helping them to prepare. Along with Charlotte's farm protocols, it was a robust induction process that Mary was proud to have put together and one she hoped would help Claire get up to speed as quickly and effectively as possible.

'We've put a lot of thought into this, Mary,' said Peter. 'Well, you and Charlotte have done most of the work to be honest, so I'm sure it will all go smoothly.'

He took his mug of tea with him back out to the parlour to see how the team were getting on, kissing Mary on the cheek as he left.

'No need to worry, love. It's all going to be fine – you'll see.'

Mary was left alone with her thoughts. She felt the urge to speak to the Chairman as it had been a few weeks since their last conversation. She picked up her mobile phone and dialled his number. As always, after a couple of rings, the Chairman answered.

'Good morning, Mary. It's set to be a wonderful day, isn't it?'

'Good morning, David; I certainly hope so. That's the reason for my call this morning.'

'It's all in the planning and preparation though, isn't it, Mary?' said David, interrupting her thoughts. 'I'm sure if you've considered what needs to be done then you've given yourself the best chance of success.'

Mary wondered if he already knew what was happening on the farm. He always seemed to be one step ahead of her.

'Our new farm manager starts today and, yes, we have prepared well for her first few days and weeks. But I'm still slightly nervous we may have forgotten something important that may harm her new career with us.'

'Planning and preparation are key, Mary. For many, a desire to crack on with what needs to be done means they ignore the importance of planning and preparation. Many believe that throwing people in at the deep end is the best way for them to learn. While it's good to be challenged, this approach often means they sink rather than swim! Sticking with the metaphor, it's far better to throw people into the shallow end where they can stand up and look around before they start to swim. It's important for people to find their feet as this builds confidence. You see, Mary, plans can be changed but the absence of a plan means change is almost certain to be necessary.'

Mary told David about the plan for Claire's first few weeks on the farm. It was quite detailed for the first week and then became less detailed as the days went on. There were clear targets for Claire to achieve and regular updates and meetings planned to ensure she was integrating smoothly with the farm and the team. David complimented Mary on her plan, which she attributed to the help of James Glover.

'James is a very wise man,' said David. 'It would be good for you and your team to follow his direction closely. And remember, planning and preparation is the quickest way to real success. Go well, Mary.'

As often happened, Mary sat and pondered what they'd spoken about. She was fascinated by the fact that he always answered her call seeming to know what she was going to talk about and ended the call when he felt he'd said enough. He was never engaged, never sounded as if she'd interrupted him and was always ready to listen. He had reminded her again in this latest call about listening to herself.

She headed to the office to find her notepad, flipped open the pages and read through the notes she'd already taken from her calls with David.

Find the silence.
Recruit in haste – regret at your leisure.
Read between the lines.
Trust your intuition.
A breakdown is often a breakthrough.
Watch for the sequences.
Put family first.
Our job as leaders isn't to try and change people but to help them find their own strengths.
Your people don't want feedback; they want your dedicated attention.
Follow the nudges.

She decided to add a couple more notes.

Plans can be changed but the absence of a plan means change is certain. Planning and preparation is the quickest way to real success.

'Morning, Mum,' said Charlotte as she entered the farm office. 'Are you excited about today?'

Mary closed her notepad and turned to face her daughter. 'Absolutely, Charlotte. It feels like the dawning of a new era for us all, doesn't it? We might be putting too much expectation on Claire's shoulders but I can't help but feel that she's not just an extra pair of hands – she'll also help us with leadership and management too. It's going to be good for all of us.'

'I feel the same way, Mum. I was talking to Martyn last night and telling him all about our plans for today and the staff meeting. I could hardly contain my enthusiasm and I didn't sleep well last night either. Is it wrong to have such high expectations, Mum? Claire isn't a superwoman who's coming in to save us all, but it does feel a bit like the Messiah is coming!'

They both laughed and continued to discuss Claire's first day and in particular the staff meeting. Charlotte had prepared a few slides they were going to print off and share with the team about the performance of the farm. They'd already agreed targets with Claire on fertility rates and a tightening of the calving block. Charlotte had spoken to the team over the past few days to find out how they were feeling and without exception everyone, including Andrew and Billy, was looking forward to Claire's arrival. It was as if she'd made an impact on the team even before she'd arrived.

'So how was Martyn when you spoke to him last night?' asked Mary.

'Well, he didn't seem his usual bright self but he hasn't been the last few times I've spoken to him. When he popped in to see us last weekend, I thought he'd lost weight, and when I challenged him about whether he was looking after himself, he said he was often skipping meals as he couldn't be bothered to cook for himself. I'm a bit worried about him, Mum.'

'Your dad and I are worried about him too,' said Mary. 'He left here feeling so excited to be driving the new tractor and getting stuck into a new role – but it doesn't seem to have worked out in the way he'd hoped. The problem is he's too proud to say so and admit that maybe he was a bit hasty in taking the decision to move. So much has happened here since he decided to leave. Do you think Martyn would ever consider coming back to the farm?'

'I did ask him last night but he seems determined to make his new job work,' said Charlotte. He told me Barry had promised to get his house sorted in the next couple of months and he was feeling a little bit more comfortable with the team. I didn't believe him, though. The bubbly Martyn I know just wasn't there. But I guess he's a big boy now and has to make his own decisions. I'd love him to return to the farm but he thinks he's just going to continue to fall out with Dad because he won't change things or listen to him.'

'But things are already changing and for the good of everyone,' said Mary. 'There's been more change on this farm in the past few months than in the past 20 years. It may only be one person joining the team but in your dad's mind, there's been a seismic shift in his thinking. He's even been talking about taking us on holiday! If your dad can change his thinking so much, I'm certain he and Martyn could adjust their thinking towards each other too. Do you remember when James did the VITA Profiling session, he talked about the importance of everyone flexing their personality style towards those whose style may be different to theirs? Perhaps Martyn and your dad could do this too?'

'Miracles do happen, Mum,' said Charlotte.

They both laughed and hoped that one day Martyn would return and accept the differences between himself and his dad. They both hoped Peter would be able to reconcile his differences with Martyn as well.

28

As Peter drove the short distance down the lanes to meet Andrew, he thought about the day ahead. Part of him felt he was handing over the farm too soon. He'd always imagined he would pass the farm to Charlotte or Martyn, or hopefully both of them together. But today, he was beginning the process of handing over the day-to-day running of the farm to Claire Davis.

It felt a bit like the start of his 'retirement', which had always been a word that made him shiver. Farming wasn't just his job, it was his life, so he wasn't sure what else he'd do if he wasn't working on the farm every day. Despite Mary having suggested many times he should have an interest outside the farm, Peter always said the farm *was* his interest. With the prospect of Claire taking over some of the decision making, direction and daily tasks on the farm, he felt a heady mix of relief and fear. Peter knew Claire's arrival would ease the pressure on him and Charlotte in particular but it also placed another prospect in front of him that he would have less to do. This certainly didn't excite him.

Peter turned the Land Rover into the field where Andrew was gathering the sheep and jumped out. 'Morning, Andrew, how are you today?'

Andrew was whistling commands to Bertie, who was busy herding the sheep towards a pen ready for shearing. It was going to be a busy time, although Andrew enjoyed seeing the fleeces taken off the farm and sold, leaving the sheep looking bare but fit and healthy.

'Morning, boss. I'm good, thank you. How are you feeling about today then with a new skipper in town?'

Peter remembered the strict instructions from Mary to be positive with all the staff about Claire's arrival. She'd said it was important they knew he and Mary were fully behind this appointment. Any doubt expressed would transfer to the team and make Claire's first few days and weeks in the job more difficult.

'I'm excited, Andrew. Today is the beginning of the next stage of the farm's development, potentially ahead of Charlotte and Martyn taking over the farm in the years to come.'

'I was talking to Charlotte the other day,' said Andrew. 'She seems really excited about Claire's arrival and it's as if they're great mates already!'

'Yes, they enjoyed a good conversation when they first met and they've kept in touch since. Claire has asked for lots of information about the farm, the cows and the key performance indicators, which forced Charlotte to gather the information and pass it over to Claire. It's helping to build a bond and understanding even before Claire starts.'

'That's great news, but most importantly, I think it's such a good move for you and Mary. It might even give you a little bit of breathing space to take time off, eh? That's not happened too often, has it?'

'Mary has already sorted that out, Andrew. We're going away in a few months' time for a couple of weeks. We wanted to make sure Claire had settled in first but Mary has taken no time at all to book a holiday. It'll be the first time in many years we've been away for two weeks.'

Andrew was pleased for his boss and close friend. Over the years, he'd watched Peter work seven days a week, often up to 12 to 14 hours a day during busy times. While Andrew himself was always busy looking after the sheep, between him and Billy, they'd always managed to get weekends off and time on holiday. When Charlotte and Martyn were young, Peter often skipped holidays. While Andrew loved the farm and often said he needed nothing more, he also loved shooting a few times a year as well as painting. Over the years, he'd attended classes, with his better pieces sold in local shops as 'paintings by a local artist'.

'I'm glad Mary has finally dragged you away from the farm. Hopefully, you'll have time to take up hobbies as well. You know I love the farm but having a hobby or two does help you to refresh your appetite for the work.'

Andrew continued to whistle commands at Bertie, who was busy running backwards and forwards bringing sheep towards the pen Andrew had prepared.

'All I need now is for Martyn to come back to the farm,' said Peter.

'How's he getting on at Burton Farm?' asked Andrew. 'Last time I asked Charlotte she said he was enjoying the new equipment but it might not be as well run as our farm.'

'Yes, you're right. I've heard a few people mention Barry isn't bothered about having a tidy farm and gets through equipment quite rapidly. Even though Martyn can live with elements of chaos, far more than I ever could, I think even *he* is finding the lack of organisation a little frustrating. However, like me, he's also a bit stubborn so doesn't want to give in just yet. I think he might feel a bit embarrassed that he took the first job on a farm he went to visit.'

Andrew leant on his shepherd's crook and looked towards Peter. 'It's interesting how some seem to learn the hard way that the grass isn't always greener on the other side. When I was younger, I occasionally wondered if there might be a bigger or better opportunity for me beyond Wilson Farm. But when I sat back and considered what I had, I realised that working in a job I loved with people I respected in such a beautiful location takes a lot of beating. That's why I've stayed here, Peter.'

'Well, I'm glad you've stuck with us, Andrew. I don't tell you often enough but I really do appreciate everything you do for us.'

'Thanks. I really appreciate that. Appreciation goes both ways, my friend. Over the years, you've treated me more like a brother than an employee and that's another huge reason why I've stayed.'

Andrew had never married and at times he felt lonely but Peter was always there to listen, talk and have a cup of coffee or just a walk around the farm. It was those moments of genuine interest from Peter that had sustained Andrew over the years.

'You're as much part of the family now as Charlotte and Martyn,' said Peter.

'Thanks. That means a lot.'

'I'll leave you to it then, Andrew, as it looks as if you've got everything under control here. When is Ben Jones coming to do the shearing?'

'He's here later this morning and depending on how he gets on, he'll be here tomorrow as well.'

'Just give me a call if you need any help or support.'

Peter headed back to the Land Rover, climbed in and sat for a few moments watching Andrew and Bertie herd the sheep towards the pen. Andrew had been in his life for many years. While Mary was his long-time companion and an amazing support, Andrew was indeed like the brother he'd longed for.

He recalled a conversation with Archie Johnston, one of the oldest members of his discussion group, who told his fellow farmers that if they really wanted long-serving employees, they should treat them like family. Archie never understood why farmers would provide poor accommodation for their staff or dangerous working conditions. He told the group that if they wanted people to stay on their farm, not to give them accommodation or working conditions they wouldn't expect their own families to work or live in.

The average service of an employee on Archie's farm was ten years. He would willingly take on younger people who were at the beginning of their careers and develop and train them until they were ready to move to other farms. This combination of youth and experience worked well for Archie and he always ensured his team time took time off for important family events. He was known for having the best accommodation for farm workers anywhere in the local area. He'd taken pride in his staff as well as his farm, telling everybody the secret to success lay in finding and keeping good people. With Claire joining the team, Peter fully intended to treat Claire as one of his family too.

29

Martyn didn't think he could spend another day at Burton Farm. The dream of a new job, a new tractor to drive and a chance to implement his ideas had turned into a nightmare.

Today had been the last straw. He'd received a telephone call at 4.30 am asking him to come into work even though it was his day off. Barry hadn't organised the rotas properly: he'd forgotten one of the staff members was on holiday so there was no one to milk the cows. While Martyn was always prepared to do the milking as it was part of his role, he was primarily interested in the machinery side of the farm. He'd rarely sat in the gleaming cab of the new tractor, which was being driven primarily by Barry, leaving Martyn to battle with various old tractors that had seen much better days.

The chaotic nature of the way the farm was run made Martyn realise just how efficiently Wilson Farm was organised. For the first time, Martyn could see the value of the consistent approach his father and grandfather had taken.

It was a warm day and there was a stale odour of damp in his mobile home. He sat at a small table in the living area with a blank piece of paper in front of him, intending to write out his notice letter. He had no idea what he'd do next but did know that he couldn't spend too many more days on this farm. It was too depressing. He hadn't spoken to his dad to see if there was a job back on the family farm but he did intend to let them know he'd resigned from Burton Farm. He hoped they might take him back but as they'd now installed a farm manager, he realised that they may not need him.

Martyn began to write. He'd never written a resignation letter before, so simply addressed the top of the letter to Barry at Burton Farm.

I hereby inform you of my intention to resign from my position as general farm worker at Burton Farm with immediate effect. Under the terms of my contract, I must give you one week's notice, which means my last day at Burton Farm will be June 15th.

Yours sincerely,
Martyn Wilson

Other members of the team who had recently resigned said that Barry had tried to talk them out of it but Martyn knew no amount of extra money or reassurances from Barry could improve the situation at the farm. He'd been promised so much and yet the job had delivered so little. In hindsight, Martyn knew he'd been attracted by the shiny new tractor and the possibility of implementing the ideas his dad refused to consider. In a reflective mood, he realised that the years of experience his dad and grandad had enjoyed on the farm meant that many of the ideas Martyn suggested had been tried before and for various reasons hadn't worked. This didn't mean his ideas were bad but he accepted that not all of them could have been implemented. He also remembered something James Glover had said. If he'd given his ideas a little more thought, supplying the detail his dad often needed and desired, his ideas might have worked. For the first time, Martyn felt that he understood his dad. He could see how different they were but that together they could create a powerful combination of challenge, new ideas and considered decision making.

'You're an idiot,' Martyn thought to himself. 'If you'd just stayed on the farm where you were, you could've worked things out with Dad.' Martyn folded up the resignation letter, put it in an envelope and placed it in his pocket. As he left his mobile home and stepped into the warm sunshine of a new day, Martyn wondered what was going to happen next. He had no real plans other than to hand in his notice and get away from Burton Farm as soon as he could. He walked up the short track behind the silage pits towards the farm office, where he hoped to find Barry.

30

Mary was in the office, paying suppliers. Her mobile phone started to vibrate and she saw it was David's number. Whenever they'd spoken, it was Mary who had called the Chairman. She looked at the phone and wondered why he was calling. Her tummy turned over with a sense that she was about to have a very important conversation. She picked up the phone.

'Hello, David, how lovely to hear from you.'

'Hello Mary, how are you today? It's a glorious day, isn't it?'

'Yes, it's a beautiful day here, thank you David. What can I do for you?'

'I just wanted to talk to you about unexpected events. We can find ourselves needing to alter plans and adjusting our goals and dreams because circumstances suddenly change.'

'Yes, I guess so, David,' said Mary. However, she still wasn't sure why the Chairman had called or the point he was making.

'How has Claire been handling her first few days?' he asked.

'Claire is doing well thanks, David. The team have really taken to her and her partner. They already feel like family. She's involving the team in developing the solutions we need. She doesn't just tell them what to do, she asks them for their views and opinions on how best to handle things so that there are no surprises.'

'Exactly, Mary. How we deal with the unexpected events or surprises in our lives often defines us. Many of us prefer to hold on to the way things have been done before and when unexpected events occur, we don't have a map of how to navigate the new territory. But there is always a way to move forward, even with dark, difficult or unfamiliar situations.'

Mary agreed but still felt puzzled.

'Remember, Mary, the unexpected events are often the catalyst for a change of direction or a new phase in our lives, just as it was with Peter's heart attack. But the next change could be just around the corner. We just need to embrace the chance to maximise the opportunity that exists within every difficulty, change or crisis we face. Have a lovely day, Mary. Speak soon and go well.'

Once more, the phone clicked and David was gone. Mary sat staring at her phone for a few more moments, reflecting on what David's message was about. She reached for her notepad.

Find the silence.
Recruit in haste – regret at your leisure.
Read between the lines.
Trust your intuition.
A breakdown is often a breakthrough.
Watch for the sequences.
Put family first.
Our job as leaders isn't to try and change people but to help them find their own strengths.
Your people don't want feedback; they want your dedicated attention.
Follow the nudges.
Plans can be changed but the absence of a plan means change is certain.
Planning and preparation is the quickest way to real success.

She added two more notes:

Expect the unexpected.
It's how we respond to the unexpected that creates the new opportunities ahead of us.

Mary returned to her computer to make the next batch of payments. She smiled, as the very next payment on her list was to James Glover from REAL Success for the Team Dynamics workshop and the recruitment of Claire. She continued to pay invoices and was deep in concentration when she heard a voice behind her.

'Hello, Mum.'

Mary spun round in her chair. Martyn was standing in the doorway. She leapt from her chair and threw her arms around her son.

'What are you doing here? We weren't expecting you today! Is everything OK? Are you all right? Are you unwell?'

'Slow down, Mum, I'm fine.'

'So why didn't you tell us you were coming? I'd have made cake or organised lunch if I knew you were paying us a visit today.'

'It might be more than a visit, Mum.'

'Oh no, you haven't lost your job, have you?'

'No, Mum, I've quit. I handed in my notice this morning and I finish next Friday.'

Mary waved at the chairs. 'Let's sit down; my knees are wobbling anyway from this surprise.'

'So, what on earth has happened? What's gone wrong? Have you fallen out with Barry?'

'Nothing has gone wrong, Mum – it's just that the job hasn't turned out to be what Barry promised at the interview.'

'But surely you've spoken to him about this? You said he was an approachable and friendly chap?'

'He's approachable, yes – but totally disorganised. Since my first day, I've barely spoken to him and there's been no review with me to check in about how things have been going.'

Mary felt frustrated, angry and upset for Martyn. This was meant to be the fresh start he wanted to kick on his farming career.

'That's ridiculous,' said Mary. 'I can't believe he hasn't spoken to you to see how you are!'

'But it's also the broken promises, Mum. I was told there would be a two-bedroom cottage but, as you know, I've been living in a damp mobile home for the past month. He told me I could drive the new tractor, which was a big draw for me, but I've barely set foot inside the cab.'

'Surely you need to give it a bit longer, though?' said Mary. 'James Glover talks about the first 100 days being the most important in a new job and you've only completed about 30. Surely the job will get better in time?'

'I've thought about this night after night,' said Martyn. 'I've looked around the farm and met the rest of the team. It's as if Barry has deliberately employed people who are just like him. I think in some ways I'm like him too but I've realised it's our standards that are different. We have high standards at Wilson Farm and I can see that now. I've also thought about Dad and our differences. These could have been the strongest part of our relationship if we'd just made our differences work. I remember this was what James Glover said at the Team Dynamics workshop. At Burton Farm, there are no differences – everyone is the same and I can see now this is a fundamental weakness, not a strength.'

Mary was reeling from dealing with the excitement of seeing her son, the unexpected nature of his visit, the call from David that preceded Martyn's entry into the room and the idea that Martyn may have realised his dad was just different to him.

'I'm struggling to take all this in if I'm honest, Martyn. So why are you here?'

'I want to come home,' said Martyn.

Martyn looked down as he didn't want his mum to see the tears forcing their way out. He took a breath and looked up. 'I've missed you all so much. I thought I was ready to do my own thing, live by myself and start a new life, and while I do still want all those things, I think I can achieve most of that here. Yes, I need to have a conversation with Dad about him giving me a little bit more flexibility and freedom but I now respect, more than ever, his approach to the farm and how he's been running it for all these years.'

Martyn was struggling to hold on to his emotion and his voice cracked. 'Will you have me back please, Mum?'

Mary was overcome with joy and threw her arms around her son. 'Of course we will! Your dad said there would always be a place for you here. Wilson Farm is your farm too. It's where you belong.'

'So how are we going to tell Dad?' asked Martyn. 'Do you think he'll be angry with me?'

'Angry? No, not at all,' said Mary. 'Your dad has changed his mind about several things over the past few months. Claire has been an absolute revelation as he can trust her to manage the farm for him. He's coming in for meals with me now. We've even taken time off and have a holiday planned! We went for a walk together last weekend and he didn't check his phone once.'

'How does Charlotte feel about Claire? Is she still feeling positive about her now that she's here and taking everything on?'

'Absolutely,' said Mary. 'In fact, Charlotte, Annabel and Claire have become really good friends. Charlotte spends more time with them than she does here with us! It's as if they're all sisters together. Charlotte also talked to us about the stress and anxiety she'd been feeling. We've talked many times since then and she's enjoying her job more now than she ever has in the past because she feels less weight of responsibility. She still hopes to take on the farm one day but wanted to do so with you. Charlotte will be so excited to see you back as well.'

They talked more about how they should approach the rest of the family. Martyn's biggest fear was how his dad would respond to the possibility of him coming home. They'd clashed so much and so often that Martyn was worried the relationship was damaged beyond repair.

'What on earth are you doing here?' asked Peter, entering the farm office and walking towards Martyn. Peter hugged his son and held him tightly.

'I can't believe you're here, son. It's so good to see you.'

As father and son held each other in a tight embrace, neither wanted to be the first to let go. Peter broke the silence.

'What's happened? Is everything OK?'

'I've quit my job at Burton Farm, Dad.'

Peter pulled back and held Martyn's shoulders.

'Come on, son. Sit down and tell me all about it.'

Martyn sat down on the remaining office chair and Peter perched on the side of the desk.

'To be honest, Dad, Barry is a bit of an idiot.'

They all laughed, releasing some of the built-up emotion between them.

'I could've told you that,' said Peter. 'But I don't think you were ready to hear it. I've met the chap a few times and he's nice enough but everything I hear about his farm sounds like a chaotic mess.'

'You can say that again, Dad. It's not only chaotic but I also think it's badly run and there's not enough care and attention about the welfare of the animals. I've just not been able to do the job I thought I was employed to do. And you remember the lovely new tractor I told you about? I've hardly stepped foot inside it.'

'I'm so sorry to hear this,' said Peter. 'I really hoped it was going to be the change you wanted. And I know the new tractor was a big draw for you so it's a shame you've not been able to use it. Perhaps it's time we replaced ours, then?'

Martyn laughed. 'Since when have you been interested in new tractors, Dad?'

‘Oh, since Claire arrived I’ve been looking at the farm in a whole new way. I’ve had time to step back and think. I’ve also realised just how much I’ve missed having you around.’

‘That’s great to hear, Dad, but I guess it took a heart scare for you to realise you needed to let go, yes?’

‘You could be right, Martyn, but I think everything happens for a reason. Your mum often says events seem to occur in a sort of order. Maybe you needed to go down to Burton Farm to see how another business is run to appreciate what we do here?’

‘You’re so right. The way you and Mum run the farm, and Gramps ran the farm before you, I can see now how it just… works. Maybe there *is* modernisation that could help us but we still need to run the farm in a structured way. I value and can see this so clearly now.’

‘You mean I’m right and I’ve been right all along?’ said Peter with a beaming smile.

Mary interjected. ‘No one needs to be right or wrong here, dear. It’s really what happens next that we need to talk about, isn’t it?’

‘If I accept you were right all along, Dad, can I come back and work on the farm?’

Peter grinned. ‘You don’t need to do anything to get your job back, son. It’s always been there, ready for you when you returned. There’s nothing you need to do or say; the job is yours when you want it. Charlotte will be overjoyed to have you back because she’s missed you too.’

‘Do you think so? Is she not going to say “I told you so” to me a hundred times?’

‘Probably,’ said Mary. ‘You’ll just have to suck that up for a few weeks. But it’s easy to look back on our decisions with regret. The older you get, the more decisions there are to regret but it’s how we respond to the events in front of us that’s much more important. So let’s move past any whys and wherefores and focus on what’s ahead for all of us and have a lovely cup of tea to celebrate.’

‘I think I might need something stronger,’ said Peter.

‘I think it’s too early to celebrate, Dad. I need to find out how Charlotte feels and Claire too. She might feel that if I come back to the farm full time, I may be pushing her out a bit.’

‘I don’t think so, Martyn,’ said Mary. ‘Claire has already worked out what you could do if you returned to the farm. She predicted you wouldn't find another farm that was run as well as ours and would come back. She was right, wasn't she?'

'It seems that way, Mum,' said Martyn.

'It will mean driving our old tractors for the time being,' said Peter.

'Those tractors might be old, Dad, but they're familiar. And that somehow feels quite comforting right now.'

The door into the kitchen burst open and Charlotte appeared, looking flustered and upset. She stopped in her tracks when she saw Martyn.

'What the heck are you doing here?'

'Hi, sis,' said Martyn.

'More importantly right now, Charlotte, what's up?' asked Peter.

'No, even more importantly, what the heck is Martyn doing here? Have you been fired already, brother?'

'No, Charlotte, I've resigned. And I'm coming home.'

She paused and walked towards Martyn, who stood up to greet her. She threw her arms around him and her shoulders began to shake as she burst into tears.

'Gosh, sister, I didn't know you cared so much.'

Charlotte didn't respond but continued to cry, with her head buried into her brother's neck.

Martyn looked at his mum and dad with a puzzled expression as he wasn't used to such a show of emotion from his sister.

'Dottie has died,' said Charlotte through her sobs. She let go of Martyn's embrace and turned to her mum and dad.

'Dottie is dead,' she said once more.

Mary moved and put her hand on her daughter's arm. 'What happened, love?'

'I went to the shed to see how she was and found her lying on the floor, not breathing. She's gone, Mum.'

'She wasn't well last night,' said Peter. 'I gave her a jab of antibiotic, hoping it would give her a boost, but her breathing was already laboured. I was worried about her, to be honest.'

'Why didn't you tell me last night, Dad? I would've spent the night with her. You know how important she is to me!'

Mary put her arms around Charlotte. 'Sit down, love.'

As Charlotte sat down, everybody else joined her so the four of them were together once again. They sat in silence for a few moments.

'She's enjoyed an amazing life, Charlotte,' said Martyn. 'You've taken care of her from day one. How many calves has she produced?'

'Last year was the seventh lactation,' said Charlotte. 'Yes, I know she did have a good life and we have taken care of her but after such a tough start in life, I wanted her ending to be so much more comfortable. It's cruel that she died alone.'

'Endings and beginnings are all part of the circle of life though,' said Mary. 'Today, Martyn has come back to the farm and it's time to start a new chapter with the family back together. Yes, it's the end of Dottie's life but life on the farm will go on, Charlotte, through all the cows in the herd that have inherited her genes. It's what farming is about. It's an endless line of endings and beginnings, isn't it? We see birth and death every single day and no day passes without something happening to make us happy and sometimes sad. Seeing the four of us around this kitchen table has made me realise nothing is more important to me than my family.'

Peter reached over and held Mary's hand. He then reached for Charlotte's hand and Martyn reached out and joined their hands with his.

'Despite our differences, we are so lucky to have each other,' said Peter. 'What makes this family so special is that we live every day with life and death. We faced the prospect of my death together too. But as Mum says, sometimes those events we see as an ending or a breakdown are the beginning of something else or perhaps a breakthrough of some kind. I've realised this so much over the past few months. Dottie's passing is sad but inevitable. Just like change, we can't prevent it, so we just need to embrace it.'

'Blimey, you two,' said Martyn. 'Since when did you both become so philosophical?'

Everyone laughed and Charlotte's sobs began to subside. 'I guess you're both right,' said Charlotte. 'Endings and beginnings are all part of life. And brother, I can't tell you how happy I am to see you back. There's been an empty chair around our kitchen table for far too long. I can't wait to tell Claire you're coming home.'

'Thank you, sis. You drive me crazy at times but oh my goodness I've missed you too.'

Peter was overcome with emotion, pride and joy to see his family back together.

'So what happens now?' asked Martyn.

'You go back and work your notice period at Burton Farm,' said Mary.

'We'll get the team ready for your return in a few weeks' time,' added Peter.

'Do I really have to go back?' said Martyn.

'Yes, Martyn,' said Mary. 'It's about being respectful to your previous employer, however difficult it's been. You can come back on your days off if you wish.'

The family talked a little more about Martyn's return and the sort of role he could fulfil. It was clear that since he'd left, nobody had been as happy on the tractors as Martyn was. Peter had picked up most of the work Martyn had been doing and they all agreed that Martyn's return would enable Peter to step back even further from the daily management of the farm.

'Maybe we can relax even more on our holiday now,' said Mary.

After hugs and gentle goodbyes, Martyn drove back to Burton Farm to finish his notice period. Peter and Charlotte headed over to the cowshed to make Dottie more comfortable. Mary stayed in the farm office to finish her paperwork. The Wilson family were a united force once again.

31

As Mary sat at her desk, she felt the urge to ring the Chairman. She dialled his number and he answered, as ever, within a few rings.

'Hello again, Mary.'

'Hi, David. Can I ask if you knew Martyn was going to return to the farm?'

'Life is full of unexpected twists, turns and surprises that we don't understand, Mary. I've said to you before that life often only makes sense when we look back. It's a strange paradox. Even during the difficult events in life such as the times when we lose someone or something, we often gain something else. It can be hard, testing and confusing at times but hidden deep within every difficulty is the seed of an opportunity to move forward. Your life, and the lives of your family and the people around you, are changing every day. Life is a constant swell of accepting the new and letting go of the old.'

Mary wasn't sure how to respond. Throughout the past few months, she'd either felt the nudge to call him or he called anyway, seemingly out of the blue, at just the right time. Even though they'd never met, she'd grown fond of him and the comforting sound of his voice.

'Thank you for being there at the right times, David,' said Mary. 'You always seemed to know what was happening on our farm and at times I wondered whether you were watching our every move.'

'Every move we make and every action we take is noticed, Mary. Over the past few months, you've seen that it's the people in our lives who matter most to us and if we're lucky enough to have a family, then that's what is often most important to us. You see, Mary, it's other human beings who enrich our lives, who can help us when we're in need and who can make each day a little easier. We must value the people in our lives but also in our businesses too. People are the ones who make our lives better and easier – even if at times human beings can be frustrating and demanding. Happy people in your team will create a happy farm, Mary, and this will translate directly into the success of your business. Go well, Mary.'

She heard a click and the Chairman was gone. Mary wondered once more about the enigmatic figure who'd been there throughout the difficult times in the past few months. She pulled out her notebook to make what she felt might be the final note to herself from her conversation with the Chairman.

Find the silence.
Recruit in haste – regret at your leisure.
Read between the lines.
Trust your intuition.
A breakdown is often a breakthrough.
Watch for the sequences.
Put family first.
Our job as leaders isn't to try and change people but to help them find their own strengths.
Your people don't want feedback; they want your dedicated attention.
Follow the nudges.
Plans can be changed but the absence of a plan means change is certain.
Planning and preparation is the quickest way to real success.
Expect the unexpected.
It's how we respond to the unexpected that creates the new opportunities ahead of us.
Happy team, happy farm.

As she sat alone with the sounds of the farm making a gentle hum in the background, Mary realised that everything that had happened to them over the past few months had been taking them forwards, even if at times they felt as if they were going backwards. David had mentioned the paradox of the breakdown often being a breakthrough and Mary could see how this directly applied to her family. She vowed to remember that people, family and the relationships between them, including their differences, would always be at the heart of Wilson Farm.

She decided to ask herself a question, just as the Chairman had told her to do when she needed an important answer.

'Who are you, Chairman?' asked Mary.

'I'm whoever you need me to be,' answered the voice in her head.

Epilogue

Six months later

'Come on, Dad, jump into the buddy seat.'

Peter climbed into the cab and sat beside Martyn in the big green tractor.

'She's a beauty, there's no doubt about it, son. You'll have to show me how to use those controls. It looks like a spaceship in here.'

Martyn laughed. 'It's easy enough, Dad. We'll get Carl Simpson to run through it with us. He should be here soon. He said he would fully demo it to us before we started using it. The software is cool and will talk directly to our farm software too, so it will save a whole load of that pesky admin you know I hate!'

'Nothing wrong with good old paper and pen,' said Peter, 'but I can see this beast of a machine is going to help us no end. I'm glad I decided it was a good idea to invest.'

Peter and Martyn laughed together and it felt good.

'Blimey you two, you look like children in a sweet shop,' said Mary, walking over to the tractor.

'Dad was right to test a few before we chose this one, Mum. I'd have gone for the first one we trialled but Dad did some research and it was clear that this was the right model for our farm.'

'Look at you!' said Charlotte, who'd appeared in the yard. 'Anyone would think farming was just about tractors and machines. I just can't see the attraction at all.'

'How's Daisy then, Charlotte?' asked Martyn.

'She's doing well. Her mum Dottie would be so proud of her. I know she was weak at first but Amy and I have looked after her together. We're quite a team now. We're very different but we complement each other so well and I love her caring approach. It's really helped Daisy in the first few months of her life.'

Mary walked back towards the farmhouse, leaving her family to talk about the big green tractor and Daisy the cow. She thought back to the time when her family was divided, her team was in danger of falling apart and her husband nearly lost his life.

She'd kept in touch with James Glover from REAL Success and he'd continued to support them during Claire's first six months in the role. They now had job descriptions for every member of the team, fully compliant employment contracts and when Marek decided to return to Poland to be with his family, James had helped them recruit a new herdsperson. Jon Knowles had slotted into the team seamlessly and his Investigator/Team Maker profile saw him get on well with everyone.

The Chairman no longer called Mary and she didn't feel the need to call him either. The family and the team were the happiest and most productive they had ever been. The milk yield had risen, the cows were strong and healthy and the farm was more profitable than at any point in its history. She sensed that finally she had happy people and as a result, a happy farm.

She smiled and recalled many of the Chairman's more philosophical pieces of advice that complemented the practical support provided by James from REAL Success.

'Thanks, David,' she whispered.

'Go well,' said a voice in her head.

The key messages

We are all different. Within a family and a team, our different personality styles can result in periods when we clash, fall out or simply become frustrated with each other. This reduces morale and individual performance and increases staff turnover. If we *are* running a livestock farm as depicted in this story, this tension can directly affect the performance and wellbeing of the animals too.

However, getting the team dynamics right can result in a strong joint purpose and the achievement of goals that were once thought impossible. The right blend of staff – either within a family or a bigger team – is essential for sustainable business success and the creation of a harmonious, high-performance team.

When the story opens the differences between Peter and his son Martyn are creating tension between them. Peter prefers to work in a structured and ordered manner, with well-developed plans, procedures and protocols. In the VITA Profiling® system, this is referred to as the Investigator personality style. He isn't keen on change and is determined to operate the farm in the way he and his father had done for many years.

Martyn, however, prefers variety in his work, enjoys change and is often full of new ideas, many of which he doesn't think through but is still determined to implement. He gets bored easily and lacks focus at times. Sociable and friendly too, Martyn has all the traits of the Adventurer personality style.

Mary is uncomfortable with conflict and prefers harmony, calmness and strong relationships between family and team. Seeing her husband and son at war is distressing for her. As a Team Maker in the VITA Profiling system, she simply wants everyone to be happy and sacrifices her own needs to ensure that others are happy and successful. Often the glue that holds the team together, Mary works hard to keep both the family and her team going.

Charlotte is a driven, focused and results-oriented person who is determined to achieve her goals. This is the Visionary personality style. She wants the team to crack on with what needs to be done and ideally to do things 'her way'. Others see her as blunt or even mildly aggressive at times. She prefers to keep discussion to a minimum, particularly if those discussions are of a highly personal nature, and gets frustrated with the more sensitive members of the team. Sadly, her style is in danger of breaking up the team, including long-serving members who were beginning to doubt whether they should stay on the farm.

Within the rest of the team, further differences in personality style become clear as the story unfolds. Amy is a sensitive, sometimes forgetful but warm person who worries about getting things wrong. She has a Team Maker/Investigator combination in her personality style. This makes her a little hesitant and indecisive at times, which frustrates Charlotte.

Andrew, the longest-serving member of the team, prefers to be left to get on with his job and doesn't enjoy being micromanaged, which brings him into conflict with Charlotte. He is an Investigator/Team Maker combination, which is slightly different to Amy as she leads with the Team Maker traits, supported by the Investigator traits. Andrew's style is the other way round.

Marek is a structured and hardworking person who prefers written protocols, systems and well-developed plans. He is the Investigator personality type who works systematically through his tasks and is happy to be left to follow them. Last minute changes to plans confuse and frustrate him.

Billy is a supportive and friendly person who is prepared to flex his approach and time for the benefit of the team. He is the Team Maker type. George is an enthusiastic but at times slightly chaotic individual who is often late, gets easily distracted and can be a little unreliable. He is primarily an Adventurer.

Claire has an unusual personality style – a combination of Visionary and Team Maker styles. There are far fewer people in the population with this personality style as the combination has contradictory elements – a desire to get results, often with a direct approach, but seeking to build a collaborative team too.

Barry Burton is keen to leave everyone to get on with jobs but shows little attention to detail. His personality style is likely to be a combination of Visionary and Adventurer, which is a highly entrepreneurial style often characterised by taking on new investments or ideas, a relaxed approach to how things are done but with a big focus on getting things done.

Edward Longstaff has a very strong Visionary style and demonstrates some of the worst aspects of this personality style with his aggressive and impatient approach. Other minor characters all have clues to their personality styles in the narrative – can you spot them?

The lack of understanding of differences in personality style and how that affects how we work is at the heart of the best and worst performing teams in any business. They are also the basis of the VITA Profiling® system designed and developed by REAL Success and referred to by James Glover in the Team Dynamics session.

The VITA Profiling® system uses a simple online or paper-based tool to establish the personality style of an individual and then, collectively, as a team. It is simple to understand, quick to administer and accurate in its description of each individual. Many farming businesses use the system to help them build stronger, more harmonious and high-performing teams. If you'd like some help creating a happy team on your farm or would like to know more about VITA Profiling® for your team, contact the team at REAL Success via our website: real-success.co.uk

Other key issues

The story also tackles other areas of the farming industry that are often not discussed or addressed, such as sexism, homophobia and mental health. While the story gently acknowledges some of the issues and perceptions, if you've been affected personally or would like support, please refer to the links section for organisations that can help, or contact REAL Success.

PEOPLE ACHIEVING TOGETHER

VITA Profiling®: People achieving together

Visionaries are bold, decisive, results oriented, driven and confident. Goals and achievement are important to them as well as getting things done. They will want to lead and believe they know best. They may be blunt under stress and impatient. Avoid approaching Visionaries with too much detail or too much 'fluffy' stuff about people's feelings. Get to the point quickly and stick to the facts.

Investigators are orderly individuals who prefer to reflect and think things through. They can excel in finding logical solutions and making sense of situations. Systems, processes and accuracy are important to them and they may be critical of others when stressed. The big picture alone doesn't work for them as they want to have data and information. Be clear, calm and patient with them to get the best results.

Team Makers are caring and considerate and are often the glue that hold teams together. Patient and kind, they can be loyal and deeply feeling people that desire harmony and flow. They will avoid conflict and always stand up for the team. Don't approach them too directly or aggressively and try to begin by asking them how they feel. Be warm, friendly and interested in them.

Adventurers look to the future with optimism, inspiring others with their ideas and acceptance of change. Having fun, enjoying an experience and being positive are important to them. They can easily become bored and follow-through may be a challenge. Avoid approaching them with too much detail and give them the freedom to create and contribute ideas.

Further note: We all have our own unique blend of the four personality styles above. We tend to lead with one or two of the styles but are able to 'dial up' the styles that are lower in our profile and 'dial down' our more dominant styles. This is key if we are to flex our approach with others.

It's also important to note that under stress or pressure, we often overextend our dominant style. Therefore, the Visionary can become even more blunt and impatient, the Investigator can be more critical and stubborn, the Team Maker more worried and emotional and the Adventurer can become more flippant and forgetful.

Changing our approach to how we communicate with each personality style can have the single biggest impact on our team's performance and likelihood of retaining our best staff.

For more information about how you can learn to adapt your style, contact REAL Success at real-success.co.uk

The Chairman

This character helps Mary to think beyond traditional leadership management principles with a broader and more holistic perspective. He performs the role of a mentor for Mary and is someone she can turn to for advice when situations become more difficult or challenging. She makes notes of his key suggestions and within these thoughts are powerful messages for readers to take into their own lives or businesses.

Going deeper into ourselves and trusting our instincts can often be one of the most powerful tools we can use in business and life. From putting family first to responding to unexpected events, the Chairman's advice binds together the personal and business aspects of our lives to show that the two cannot be separated.

The identity of the Chairman is never fully revealed to allow the reader to draw their own conclusions. However, having a mentor in business or life can bring huge benefits to our mental wellbeing, our individual performance at work and our personal lives. Contact REAL Success if you'd like to discuss mentoring for you or your team.

Resources

real-success.co.uk – a one-stop service providing HR support, people management training, VITA Profiling® and recruitment services.

fcn.org.uk – the Farming Community Network, a voluntary organisation and charity that supports farmers and families within the farming community

rabi.org.uk – an award-winning national charity providing local support to the farming community across England and Wales.

yanahelp.org – support for those involved in farming and other rural businesses who are affected by stress and depression.

thedpjfoundation.co.uk – provides support and training for those in the agricultural sector to become better aware of poor mental health and its impact within our communities.

gayfarmer.co.uk – a support service for those who are dealing with sexuality and prejudice issues within farming.

sianbushellassociates.co.uk – empowering family businesses in Wales, Ireland, Scotland and England by asking the hard questions to facilitate succession planning for change and growth.

paraisoescondido.pt – the amazing hotel in Portugal where this book was created and edited.

Acknowledgements

This story began in the UK in 2015 when I was running an open workshop about communication and how to understand your team, your clients and your family. Two people in the audience approached me after the session finished and asked if I worked with the farming community. Gaynor (a farm consultant) suggested that the workshop could help her farming client (Geoff) and I was invited to attend his farm. Thirteen years later, I would write this book based on the same workshop that Gaynor had the foresight to see could help reduce some of the tensions on Geoff's farm. Gaynor – thank you for your wisdom, support and encouragement over the years since, and Geoff – thank you for trusting Gaynor and becoming more than a client. You have both become and remained good friends and I will always be grateful for the break you gave me into the farming industry.

Thank you to all the clients that have trusted me and the REAL Success team with their own teams and families. You're too numerous to mention but if you've read this book and we've worked with you since my company launched in 2009, we truly appreciate and value your business.

To my team of REAL Success partners who have sacrificed their time, supported my mission and skilfully and professionally delivered our services and products to clients across the UK – thank you. Special mentions to Emma, Becky and Kate, who were the first tranche of partners to risk their own lifestyle to help me build the business.

Alex – thank you for your help with marketing and building the brand that is REAL Success. Phil and Craig – you were instrumental in the repositioning of my business in 2022 and your skill and creativity transformed my perception of what my business could become.

In preparing this book, the team at The Right Book Company

deserve special thanks. To Sue – thank you for taking a risk on a business book in the form of a story. To Marian – thank you for your initial edit of my manuscript – your candour and advice were priceless. To Bev and Andrew – thank you for your incredible work in polishing my manuscript, and to Paul – thanks for helping me plan the launch of the book.

Berny and Glenn – thank you for providing the amazing setting at your hotel in Portugal for me to write the first draft in 2021 and edit the book in 2022.

I need to record a huge thank you to my family. Two of my children, Jess and James, currently work in my business and have encouraged me to write this book. Thank you for keeping me sane and helping me to build our business. Thank you, Amy (my eldest), for being there when we needed your HR advice. Thank you, Andy (my elder brother), for your wisdom, guidance and encouragement. I don't think I'd have got the business or this book to this point without your support.

And finally, to my wife Samantha, who has watched me tapping away on my computer on two holidays and has supported me, not only with my writing but with the birth and subsequent business journey of REAL Success. It's been really tough at times but you've never stopped believing in me and for that I'm eternally grateful.

My final acknowledgement is to every person who has attended our VITA Profiling workshops, read my articles, listened to me at events and been interested in my work. May you continue to understand yourself and your teams and in doing so, make your families and businesses a happier and more successful place to live and work.

And reader – be yourself. There is no other version of you that matters.

Go well.

Paul

About the author

Paul Harris is founder and CEO of REAL Success, a business dedicated to providing organisations with inspiration, products and services that improve teamwork, employee engagement and staff retention. He is also the creator of the VITA Profiling® system referred to in this book.

Paul's passion for people and business is reflected in his writing, speaking and consulting. His first novel, *The Man with Scarlet Socks*, was acknowledged as a meaningful self-help book for those struggling to grasp the concept of 'real success'. When Paul is not writing, he consults with business owners and leaders to help them lead and manage their teams more effectively, recruiting and retaining their best talent. Working extensively across the agricultural sector in the UK and Europe, along with a diverse range of other markets, Paul speaks to thousands of leaders each year at conferences and exhibitions. He is widely recognised as a thought leader and advocate of simple people development and leadership strategies.

Paul has been married to his wife Samantha for 37 years and they have three wonderful children.

To learn more about Paul and REAL Success, including VITA Profiling, please visit real-success.co.uk

CPSIA information can be obtained
at www.ICGtesting.com
Printed in the USA
LVHW010226310523
748327LV00011B/156